CONTENTS

PREFACE *vii*

1. ADAPTATION *1*

Natural selection and the problem of alternatives *2*
Evidence for adaptation *5*
  1. Making use of existing genetic variation *5*
  2. Using artificially produced variation *7*
  3. 'The comparative method' *9*
  4. Adaptation through 'design features' *12*
Conclusions: is all behaviour adaptive? *17*
Further reading *20*

2. OPTIMALITY *21*

The two meanings of optimality *21*
Long-term optimality *22*
Short-term optimality *24*
The link between short-term and long-term optimality *26*
The separation of short-term and long-term optimality *28*
'Rules of thumb' *31*
Costs, benefits and economics *32*
Conclusions: do animals behave optimally? *34*
Further reading *35*

3. INCLUSIVE FITNESS *37*

Aiding relatives and 'inclusive fitness' *38*
Offspring and inclusive fitness *43*
Genes, bodies and recognition *50*
Conclusions *53*
Further reading *53*

4. GENES AND INNATE BEHAVIOUR 55

'Genes for' behaviour 56
The genetics of behaviour 58
Genes, behaviour and genetic determinism 60
Innate as 'genetic' 62
Innate as 'the opposite of learnt' 62
'Developmentally fixed' does not mean 'genetic' 66
Conclusions: should we talk about 'innate' behaviour at all? 68
Further reading 69

5. COMMUNICATION 71

Why 'communication' and 'signal' should be difficult to define 72
The two meanings of 'information transfer' 75
Manipulation and the dangers of being manipulated 78
Cost, honesty and handicaps 80
Are all signals handicaps? 82
Are all signals 'honest'? 85
Conclusions: information transfer or manipulation? 87
Further reading 88

6. SEX AND SEXUAL SELECTION 89

The advantages of sex 89
Selection on males and females 92
The exploitation fallacy 93
Male rivalry and female choice 94
Theories of female choice 96
Conclusions 104
Further reading 105

7. THE MACHINERY OF BEHAVIOUR 107

What is 'instinct' and what is wrong with it? 109
The direct approach: mapping the nervous system 112
Whole animals, 'black boxes' and motivation 114
What is 'motivation' and where does it get us? 115
Neural nets and the connectionist revolution 119
Conclusions 120
Further reading 121

8. COGNITIVE ETHOLOGY 123

What is a representation? 125
Cognitive maps 127
Concept formation 131
Conclusions: do animals have cognitive representations? 134
Further reading 135

9. CONSCIOUSNESS 137

What is consciousness? 138
The problem of consciousness 139
Functions of consciousness 140
Animal suffering 144
Conclusions 146
Further reading 147

10. ANIMAL BEHAVIOUR AND HUMAN BEHAVIOUR 149

Determinism of various sorts 149
The importance of culture 153
Time lags 157
Which animals? 159
Conclusions 162
Further reading 163

EPILOGUE 165

REFERENCES 167

INDEX 177

To subjects 177
To authors 181

Preface

This edition is, like the first, intended as a guide or a companion to help unravel some of the difficulties and confusions that are sometimes encountered in the study of animal behaviour. It is written as a lifeline to students who feel confused or overwhelmed by the mass of information now available about animal behaviour. It does not pretend to cover the whole subject but rather aims to give a readable introduction to key ideas which can then be followed up in more detailed textbooks.

The book is primarily intended for students taking a course in animal behaviour as part of a biology, psychology or human sciences degree, but I hope it can be read by anyone with an interest in the behaviour of animals. References are given in the usual way at the end of the book and a separate list of key articles is given at the end of each chapter, so that anyone who wishes to gain a fuller and more comprehensive picture of the work that has been done can do so. But to make the book easier to read, the number of references mentioned in the text itself has been deliberately kept down.

Most of the problems that are discussed in this book have, if it is any comfort, trapped me at some time or another and so I write as one who has stumbled over but eventually found a way out of difficulties rather than as a superior being who has never encountered any. I have been greatly helped by various people who have been willing to discuss issues or read parts of the book in draft form, including Alan Grafen, Chris Barnard and Jenny Clee.

Marian Dawkins
Somerville College,
Oxford
March, 1994

1
ADAPTATION

Vine snake eating frog. Photograph by Paul Stewart.

'Adaptation', as it is commonly understood, refers to characteristics of living organisms – including their colour, shape, physiology and behaviour – that enable them to survive and reproduce successfully in the environments in which they live. The most widely accepted theory of how such adaptations come about is the theory of natural selection, which is basically a theory of how small changes at the genetic level are passed on to the next generation through their effects on the bodies of whole organisms. As we will see, many of the difficulties that people encounter with understanding the concept of 'adaptation' comes from the fact that it is so easy to misunderstand the connection between changes at the gene level and changes in success or failure at the level of the individual animal. The result is that more confusion has arisen over natural selection than almost anything else in the study of behaviour. This may come as a surprise to anyone who thought that the theory of natural selection was simple and straightforward. But misunderstandings and misapplications of it are to be found everywhere. Not only have people misunderstood the idea of 'adaptation' and its connection with genetic changes, but some of them have quite wrongly assumed that theories of adaptation are untestable and nothing more than 'Just so' stories (Gould & Lewontin, 1979). There have been misunderstandings about what it means to say that animals behave optimally and, as for 'inclusive fitness', it is difficult to find anyone who has not at some stage been confused about what it means. So although it may seem an odd way to start a book called *Unravelling Animal Behaviour*, a careful look at the theory of natural selection will in fact set the scene for much of what comes later.

NATURAL SELECTION AND THE PROBLEM OF ALTERNATIVES

The theory of natural selection as put forward by Charles Darwin is sometimes seen as an assertion that all animals are adapted to their environments, with characteristics which 'fit' them to their way of life. On this view, giraffes are said to have long necks to reach the tops of trees and stick insects to look like twigs to avoid being eaten by birds.

But this is not, in fact, what Darwin's theory of natural selection is all about. His theory is not about animals simply surviving and reproducing. He did not suggest that animals and plants were engaged in a private fight with their physical environments, important though it may be for them to 'battle' with heat or cold. Darwin's theory is, by contrast, about organisms surviving *better than* their competitors. He saw animals engaged in a struggle to exist and reproduce in which the best plant or animal won. It was not enough to be good at surviving. The important thing was to be better than the competition.

So, apparently simple questions such as 'Why does a giraffe have a long neck?' or 'Why do rabbits have such big ears?' are much more difficult to answer than they appear at first sight. If large ears in rabbits evolved because large-eared rabbits did better in competition with rabbits with other sorts of ears, what were these other rabbits like? And why did they not do as well? Did they have smaller ears which could not pick up distant sounds as well or did they have larger ears which got caught up in the sides of burrows? Did they have rudimentary ears or ones of a different shape altogether?

Putting the problem in this way enables us to see that any question about adaptation really involves at least two separate questions. Firstly, there is the question of what alternatives were available for natural selection to choose between. This would ideally involve knowing what genetic mutations had occurred or, to use the colourful language of geneticists, what genes were floating around in the gene pool and how many of each sort there were. Secondly, there is the question of why one alternative did better than the others. In many cases, the competitors which failed in the struggle for existence will no longer be around; but if we accept Darwin's theory, we have to postulate that, at least sometime in the past, they did once exist and that there was a reason for their demise. If there was a change in the relative numbers of genes in the gene pool, we need to know what caused this change.

The trouble with this is that we can postulate alternative kinds of animal indefinitely – pigs with wings, green rabbits, and so on. But whether these actually were the raw material of evolution is extremely difficult to determine. In other words, it is very hard to know whether they did once exist but were found wanting in the struggle for life or whether such mutations never have and never could have arisen within that species.

If they never have existed, as rabbits with little metal wheels have certainly never existed, then they are not one of the alternatives available for selection. But whereas we can confidently rule out little metal wheels as part of the rabbit alternatives for evolution and equally confidently include rabbits with slightly larger ears than average (because they exist), there is a grey area in between. Could mutations for rabbits with differently shaped ears, or three ears arise? Indeed, have they? Were these mutants once alive and were they selected against? Or have the mutations never arisen, never given the rabbits possessing them a chance to prove themselves, one way or the other, in competition with other rabbits?

The second question about adaptation can be asked only when we have somehow overcome these problems and decided what alternatives were available for selection. The question is this: *why* did one alternative do better than the others? What happened to the losers? To put it at its most crude, what did the failures in the struggle for existence die of?

Notice that this is not a question about *whether* natural selection has taken place. It is about the fates of winners and losers. Given that there were rabbits with ears of many different shapes and sizes, why did the big-eared ones have the edge over smaller-eared ones? Was it perhaps because they could hear better and so were less likely to die from predation? Or was it because they could regulate their temperature better and so were less likely to die of overheating?

To say that something is an 'adaptation' in an animal alive now implies that we can distinguish between the various hypotheses about causes of death and reproductive failure in rivals that may have existed many millions of years ago. It is like a murder mystery that took place a long time in the past, so long ago that all the witnesses have died, and we have to work out whether the victim died by the sword or by being poisoned. The only remaining clues, apart from a few fossilized bodies, are the present-day living descendants of those who managed to survive some unspecified disasters of the past.

It is from such unpromising material that we have to derive our hypotheses about adaptation and to work out what befell those that were unsuccessful. Contrary to what is sometimes believed, however, this does not mean that all we can do is to speculate with no possibility of obtaining any sort of evidence. There are in fact four well-established methods for putting the study of adaptation on a firm basis and we will now look at these. Two of them could be described as 'direct' or 'reconstruction' methods, in that they involve trying to observe death and reproductive failure in present-day animals as indications of what probably happened in the past. We either compare two existing forms of the same species (discussed under 1 below) and see which one fares better or we compare a single variety with some artificially contrived, man-made alternative (see 2 below).

The other two methods, which are 'indirect', spare us the need to actually observe killing and death and concentrate instead on the results of natural selection. If present-day animals are the result of a struggle for existence that took place in the past, it may be possible to make inferences about what was going on then by studying the descendants of those victors now. Natural selection in the past is studied through its legacy in present-day animals either through cross-species comparisons (see 3 below) or through uncovering 'design features' of structures or behaviour (see 4 below).

Evidence for adaptation

1. Making use of existing genetic variation

The most straightforward of the direct methods is to exploit the fact that, even in present-day species, not all individuals will be exactly alike. There will be variation in size, colour, and perhaps various aspects of their behaviour that will lead us to ask whether any of these characteristics enable their possessors to survive and reproduce better. The classic example of this is the colour variation in the peppered moth (*Biston betularia*). The natural occurrence of light and dark forms of this moth enabled Kettlewell (1955) to show that in industrial areas, where a sooty atmosphere meant that most of the tree-trunks were dark and not covered with lichen, the dark form was better protected against bird predators than the lighter form. The light form was, in turn, better camouflaged than the dark moths where lichen tended to cover the tree-trunks. In this case, the colour adaptation was studied through a direct study of the death of the uncamouflaged moths. Moths of different colours were experimentally released in different areas and birds were then seen to fly onto the trunks and eat insects that stood out from the background, leaving behind the ones whose colour enabled them to resemble just another patch of tree. Death took place under the eyes of the observers.

There are a number of important features which we should notice about this familiar and often-quoted example (which has incidentally been shown to be somewhat more complicated than described here). In the first place, it is known that the differences between the dark and the light forms have a genetic basis and so their survival and death can give a clear indication of the gene frequencies to be found in the next generation. Bird predators can almost be seen to remove some genes and leave others.

The same cannot be said, however, for many other cases where one kind of animal survives and another is killed, and where it would be quite easy to imagine (erroneously) that natural selection was being observed in action. In fact, one of the commonest methods used for studying adaptation is subject to the very serious objection that there is often no evidence for any differential gene survival at all. In black-headed gull colonies, for instance, the birds nesting at the centre of the colony have a higher reproductive success than those nesting at the edge because predators such as gulls, crows, and foxes find it much easier to take young and eggs from nests at the unprotected edge of the colony than from the centre where the parent gulls mount a massed attack (Patterson, 1965). But it would be quite wrong to use this as evidence for natural selection in action because there is no evidence that there are genetic differences between gulls nesting in different places. They could all be genetically similar as far as nest-site selection is concerned but nest in different places depending on what was available. In fact, the central, safer nest sites do tend to be occupied by older gulls, with the younger ones being pushed to the more dangerous outside. We cannot conclude from the fact that chicks in outside nests are more likely to be killed that one genetic form is being penalized and another favoured because nobody has yet produced evidence that there is any genetic difference between inside nesters and outside nesters.

A second point to be noticed about the peppered moth example is that, because it was a carefully conducted experiment, it was possible to be sure that predation risk was a key factor involved but the same is not true of very many other studies that claim to show adaptation. Elgar (1989) reviews a much studied field – the adaptive significance of group living in animals – and shows that many studies simply do not show what they are claimed to do. A common method is to compare large and small groups of animals or animals on their own with those of the same species in flocks or herds, and then to measure something that is thought to be closely connected with survival or reproduction, such as the amount of food eaten or speed of response to a predator. (Usually, it is something thought to be a good indicator of survival and reproductive success, rather than actual death or reproductive failure, yet another reason for being a little sceptical as the assumed connection is rarely tested.) Elgar systematically points out, however, that over and over again, these studies jump to false conclusions. Just because a solitary animal may get less to eat than one living in a group, it does not follow that living in a group is an adaptation for feeding more efficiently. It could just as well be that groups of animals congregate where there is a lot of food and that is why the animals in the group eat more – nothing to do with grouping *per se*. To a certain extent, such difficulties can be overcome by doing an experiment (see next section) and controlling for all the possible extraneous variables we can think of. If we only compare like with like (make sure that the amount of food available to grouped and solitary animals is the same, that they are the same ages and so on) we can then be more certain that we really are looking at the genuine advantages of grouping and not the spurious effects of anything else. The advantage of setting up such an experiment is that, if it is properly done, there is much more control over the different variables.

The third point to be noticed about the peppered moth example is the fact that the effects of natural selection could be 'restaged' artificially by importing alternative uncamouflaged moths from another area where they would have been camouflaged since the paler form also persists at a low frequency even in heavily polluted industrial areas. Taking animals from one geographical area with one set of selection pressures and putting them into another area with a different set of selection pressures is, in fact, one of the best ways we have of restaging evolution. In a dramatic transplant experiment, Endler (1980) took guppies from a stream in Trinidad where there was a high risk of predation and the fish were drab in colour and released them into another stream where a major guppy predator (a cichlid fish called *Crenicichla*) was not present. Major changes took place in the descendants of the released fish. Within a very short time, the relaxation of selection pressure from predation in their new stream resulted in a descendant population of guppies that were far more colourful than their parents and grandparents had been. They had many more iridescent blue, red and black patches of colour on them. Endler found that he could, in other words, shift the frequencies of genes controlling colour patterns by staging a shift in predation pressure, thus showing how colour pattern is adapted to predation. So is behaviour. The fish became less likely to school, more likely to court in the open and generally less wary than their ancestors in predator-risky streams had been (Magurran *et al.* 1992).

In the majority of cases, however, such convenient alternatives cannot be found at all. Frequently, only one variant will remain, the others having been eliminated by the very process – natural selection – that we want to study. This may be frustrating for our attempts to study adaptation but it is hardly surprising. If natural selection has been acting in the way we think it has, unsuccessful alternatives should no longer be available for us to do our comparisons with.

So, experiments of the peppered moth or guppy kind, although an ideal and very revealing way of studying adaptation, are in many cases not feasible. Two genetic alternatives may simply not be available for comparison or the restaging of an evolutionary event from the past may not be possible. So a second, somewhat different method may have to be adopted as a substitute.

2. Using artificially produced variation

Although it may not always be possible to find the required unsuccessful variants occurring naturally, it may nevertheless be possible to create them artificially. These artificial man-made 'mutants' can then be used to show how good real animals are by comparison. There is, of course, the obvious objection that the artificial variants may not precisely mimic the mutations that have, or could ever have, arisen naturally within a population, but with a little ingenuity and knowledge of the species concerned, very plausible 'mutants' can be devised.

Niko Tinbergen and his colleagues (1967), in a pioneering experiment along these lines, set out to discover why black-headed gulls remove empty eggshells from their nests. After a chick has hatched, a parent bird will pick up the shell in its beak and fly some distance from the nest with it before dropping it well away from the nest area.

Tinbergen wanted to know why the gulls do this, why, that is, gulls that remove eggshells are at an advantage compared with those that do not. This was particularly problematical because the newly hatched chicks are very vulnerable to being eaten by crows and herring gulls while the parent is in the act of taking the eggshell away. In addition, some black-headed gulls prey selectively on the wet chicks of their neighbours that have just come out of the egg. There are several possible adaptive explanations for the evidently risky behaviour of removing shells. One is that if the eggshell remained in the nest it might damage the chick. A second is that one might slip over an unhatched egg and prevent the next chick from coming out. A third explanation is that it might make it more difficult for the parent to brood the unhatched eggs and a fourth is that shells might harbour disease. Yet another explanation is that the gulls remove eggshells because the empty shells might attract the attention of predators. Notice that – in line with what we were saying earlier about adaptation – each of these hypotheses is about a different possible *cause of death* in chicks if their parents did not remove eggshells. Discovering the nature of the adaptation means discovering which of them is most likely to have resulted in the differential survival of chicks whose parents did remove eggshells compared to those whose parents left the eggshells lying around the nest, not just reiterating that natural selection has occurred.

Because all present-day black-headed gulls, as far as is known, remove eggshells, there is no way of finding out what the advantage of the behaviour is using natural mutants that do not show this behaviour. So Tinbergen devised some artificial mutants – man-made gull nests complete with eggs from which no eggshells had been removed – and compared them with other nests with eggshells removed to different distances. He found that the least successful 'mutants' were those nests with eggshells left in them or near them. Crows and herring gulls flew down and took the unhatched eggs that remained. Nests with eggshells well away from them were much less easily spotted by the predators.

All this strongly suggests that if a real mutant black-headed gull that did not remove eggshells existed in the past, its reproductive success would have been lowered because its eggs and young would be more likely to be eaten by predators, that is, the evidence favours the last of the five hypotheses listed above. Eggshell removal is therefore most probably an anti-predator adaptation, death of offspring through predation being the most likely reason why past rivals lost out, as opposed to death of offspring through disease harboured in the shells or injury from the sharp edges.

Even this method, however, is not altogether satisfactory, partly because the experiments are very difficult to do in practice and partly because the artificial 'mutants' are only an approximation to what goes on or what has gone on in nature. If we happen to have thought up the wrong mutant, the answers to our comparison could be quite misleading. Tinbergen chose to compare removing eggshells with not removing eggshells using shells that were white inside (as real gull eggs are), but a similar comparison using shells that were green and camouflaged on the inside as well as the outside might have given very different results. It is plausible to suppose that the past gull rivals had eggs that were white inside like present-day gulls and were distinguished from their successful competitors only by a certain slovenliness over keeping their nests tidy – but that is a supposition. The original struggle for existence may have been between participants that were rather different from the 'removers' and 'non-removers' that were recreated in the present day.

As well as this direct evidence, however, there is another way of studying adaptation. This is altogether a gentler approach. The two methods we will now consider do not involve actually having to see predation or starvation taking place. They assume that the death and killing occurred in the past but they do not demand a reconstruction in the present. Their raw material is the animals of the present day, the victors of those long-ago events. By looking at what makes a victor, they attempt to work out the nature of the victor's advantage. Adaptation is studied not through reconstruction, but through its effects.

3. 'The Comparative Method'

Although all methods for studying adaptation are to some extent comparative (success is always relative to something else), there is one that has come to be known as the 'Comparative Method'. 'Comparative' here means comparisons between species (or less commonly, comparisons between populations of the same species) which are living in different areas or niches and are therefore subject to different selection pressures. Now, looking at the effects of different selection pressures is precisely what we are interested in in our pursuit of adaptation, but it may not at first sight be clear how making comparisons between totally different species throws any light on it.

The rationale goes something like this. Suppose that species A lives in an environment in which there are not very many predators, whereas species B, which is closely related to it, lives in another environment where there is constant danger from predation. Differences between the behaviour of species A and species B might tell us something about the behaviour that B has evolved specifically as anti-predator adaptations. Of course, we have to be careful with such comparisons. There may be other differences between the environments of A and B besides the obvious one of predation pressure. But the more species we look at and the greater the number that behave like A when they suffer little risk of predation and like B when predation is heavy, the more confident we can be that we have correctly identified the crucial selective force.

E. Cullen (1957) made a detailed comparison between the behaviour of the kittiwake and various other closely related gull species. The importance of her study lies in the fact that there is a crucial difference between the environment of the kittiwake and that of these other species. Most gulls nest on the ground whereas kittiwakes nest on steep cliffs, out of reach of crows and herring gulls, which are, as we have already seen, dangerous predators of the eggs of ground-nesters. Sometimes even the kittiwakes themselves have to make several attempts before being able to land on their own nests, perched as they are on narrow ledges; but at least their young are virtually free from predation.

Kittiwake nests are extremely messy. The parents do not remove the empty eggshells and their nests are made highly conspicuous by their white droppings. This is very interesting confirmation of the hypothesis we had already arrived at that eggshell removal in ground-nesters is indeed an anti-predator adaptation, because all the other potential dangers such as the shells harbouring disease or injuring the chicks would seem to apply just as much in kittiwakes as in other gulls. Yet, released from the pressure of predation, eggshell removal is absent. The fact that this behaviour occurs in ground nesting species that are close relatives of the kittiwake strongly suggests that similar tendencies could arise (and perhaps have previously arisen) in the kittiwakes themselves. A kittiwake that removes eggshells, like a ground-nesting gull that does not, was a plausible candidate for selection. The non-occurrence of this behaviour in kittiwakes and its very striking occurrence in other species clearly points to the action of selection operating through predators (Cullen, 1960).

Such comparisons become even more convincing when larger numbers of species are considered. Gannets, which, like kittiwakes, nest in inaccessible places, are similar in not removing eggshells and having generally conspicuous nests. In fact, with respect to many of its nesting habits, the kittiwake is more like a gannet, to which it is only distantly related, than to other gulls which nest on the ground (Nelson, 1967).

The power of the Comparative Method is that it shows up correlations between the environments in which animals live and the behaviour which they show and thus suggests explanations for otherwise puzzling aspects of what animals do – such as flying up with empty eggshells and dropping them. But, unless applied with care, there can be major problems with it. Chief of these is that 'correlation' does not necessarily mean 'causation'. In other words, just because there may be a correlation between susceptibility to predators and removing eggshells, it does not follow that predation causes the difference between the behaviour of kittiwakes and that of other gulls. It could be some other factor, such as the temperature of their nest sites or some unsuspected disease risk peculiar to kittiwakes. One of the best ways of guarding against this is to use large numbers of species for the comparison. The more different instances we have of species that do remove eggshells and those that do not, the surer we can be that it was not just some extraneous peculiarity of one species that was responsible for the differences between species but was genuinely the one – in this case predation – that we thought it was.

But there is a trap even here. Adding more and more species to our comparison is only going to help if the species in question evolved the trait independently. If the trait evolved just once in an ancestral species and then was inherited by all the descendant species, then we should not count all the descendant species as providing additional evidence for the adaptive nature of the trait. If the trait evolved once it should only be counted once. The similarity between black-headed gulls and nightjars in their habit of removing eggshells or between kittiwakes and gannets in not doing so is impressive because these pairs of species are only distantly related to each other and yet in their attention to nest hygiene show more similarity to each other than they do to more closely related species. Kittiwakes stick out like a sore thumb from other gulls and in the one respect of their nesting hygiene they are more like gannets than they are to their closer gull relatives. It is with this 'sticking out like a sore thumb' or independent evolution of a trait in phylogenetically distantly related animals subject to similar selection pressure that comparative studies become most convincing. Consequently, the Comparative Method cannot be used unless the evolutionary relationships between animals are already known in some detail (Ridley, 1983; Harvey & Pagel, 1991). For example, to apply the Comparative Method in a truly valid way to the adaptive nature of eggshell removal, we should really establish how closely related gulls and nightjars are, and whether their common ancestor did or did not remove eggshells.

Provided the independence of evolutionary events can be established, the distribution of a behavioural trait – that is, its presence in some animals pursuing one mode of life and its absence from other, perhaps closely related animals doing something else – is one of the most important kinds of evidence we can have in our pursuit of adaptation. What we see now are the survivors in different environments, with success being defined differently when the selection pressures vary. By comparing success stories in different environments, we begin to understand what success is and what it is to be adapted to a particular way of life.

4. Adaptation through 'design features'

As we have seen, all questions about adaptation involve some sort of comparison and yet there are enormous practical difficulties in showing exactly why one kind of animal does better than another in the struggle for existence and reproduction. Some of these practical difficulties concern, as we have already seen, the problem of re-creating animals with the relevant variation in the first place. Others concern the problem of showing that the reproductive success of the individuals that we have reconstructed from the past really is lower than that of present animals, and if so, why.

There is, fortunately, a way round such practical obstacles, one that still involves a comparison but this time not with other animals. The comparison here is between animals and the best efforts of human engineers. Take the specific example of bat echo location. Insectivorous bats catch small insects on the wing by sending out brief pulses of sound and then listening to the echoes that come back to them reflected off the insects, like the sonar system of a submarine, but much, much faster. Indeed, it is the comparison with man-made sonar or radar systems that tells us about the adaptive significance of the sounds the bats make.

If a human engineer were charged with the task of designing a sonar system to detect small moving objects and plotting their position with sufficient accuracy that they could be intercepted in flight, he (or she) would:

1. Have to use very high frequency sounds. Objects only give off reasonable echoes if the sound hitting them has a shorter wavelength than their own diameter (you will not get an echo if you shout at a needle). With an average-sized moth, the sound used would have to be between 80 and 100 kHz, that is, way above the maximum frequency that humans can hear which is less than 20 kHz.

2. Have to use sound of a very high intensity. The high frequency sounds needed for bouncing off small objects do not travel well – they get absorbed by the atmosphere – so to give the sonar system a range beyond a few centimetres, the human engineer would have to use extremely loud sounds in the first place.

3. Have to send out sonar blips very frequently to keep up with the rapidly moving target.

4. Have to devise a way in which the sonar system could receive faint echoes coming back from the target and not have the sensitive reception mechanism drowned by the loud sounds that were being sent out. And so on.

The list of requirements would have to be much longer than this before the sonar would work, but it is already long enough to make the point. It would have to be a complex machine designed very carefully to do a particular job.

If we now compare such a machine with a bat, we find that bats already do most of these things. Bats, like *Myotis*, the little brown bat, emit very high frequency (50–100 kHz), very loud (60 to several thousand dynes/cm^2) sounds at over 100 times a second under some circumstances and manage to listen to over 100 echoes a second in the brief intervals between their own sounds. They have devices for not damaging their ears, for giving priority to faint echoes over louder sounds and for accurate tracking and imaging (Sales and Pye, 1974; Simmons *et al.*, 1979; Tanake *et al.*, 1992). They can decode information about the position, nature and velocity of their targets, using the information contained in the returning echoes.

If the system had been built by human engineers, we would say that that humans had 'designed' it to detect small moving objects. As it is, we can use the comparison between what humans have done (or would do) and what the bat does to study how natural selection has been operating and, indeed, to talk about how natural selection has 'designed' bat calls for their echo location function. The closer the match between a machine deliberately designed to perform a certain function and what an animal does, the more likely it becomes that we have identified the selection pressures which have been operating on the animal. In the case of the insectivorous bats, the close similarity between their echo location system and the echo-detecting abilities of human sonar systems suggests that the bats' behaviour is adapted to catching insects on the wing and not, say, to scaring off predators or communicating with other bats because neither of these alternative adaptive hypotheses would explain why the bats' calls are of such a high frequency, are so loud and are repeated so often. Death by starvation, through giving a call that was ineffective at catching food, would appear to have been the fate of the bats' now vanished rivals, not death through being eaten or death in any other guise.

This conclusion about the adaptive significance of bat echo location is, however, reached without anybody ever having done a comparison of the reproductive success between a typical little brown bat and one of the same species that made an abnormally low frequency sound or one that was abnormally quiet. No one has shown that the atypical bat would leave fewer offspring through not being able to catch so many insects. But understanding the 'design features' of the echo location system makes this unnecessary.

Catching prey is the only adaptive hypothesis that adequately explains all the unusual features of the bats' behaviour. Another possible hypothesis, that the bats which make such sounds are safer from predation than bats which make other sounds, would not explain why the sounds that are made are at a frequency far above that which any possible predators can hear. Nor would it explain why the bats appear to adjust both the length and the timing of the sounds that they make so that the echoes of their own voices always come back in the intervals between the sounds they are making. Only the hypothesis that the sounds are an adaptation to the extraordinarily difficult task of catching small insects on the wing explains the complexity of bat calls. Only this hypothesis accounts for all such details as the duration of each sound, the way the sounds get shorter as the bat homes in on its prey, the frequency composition and the curious change in frequency (a drop of about an octave) during each pulse, and so on. The close similarity to a complex piece of human machinery allows us to reach this conclusion. The 'design features' of the machine and the parallels we see in the bat suggest that natural selection has discarded those alternative forms that did not live up to these criteria in the same way that an engineer might reject prototype machines that did not do what was required of them.

The value of this method of studying adaptation can be seen even more clearly if we now look at a negative example, that is, one in which a hypothesis about adaptation has to be discarded in the face of 'design feature' evidence. It had been supposed for a number of years that one of the reasons why many fish travel in schools is that they gain a hydrodynamic advantage from doing so, that is they make use of the 'eddies' produced by other fish and so do not have to expend so much energy in swimming themselves. If fish schooling behaviour is 'designed' in this way, it should be possible to predict how the fish should position themselves to gain the greatest advantage from the swimming of their neighbours (Fig. 1). Partridge & Pitcher (1979) showed, however, that for at least three species of schooling fish, the animals did not position themselves in this way. They may have been gaining some hydrodynamic benefit, but it was certainly not as much as they theoretically could have done by choosing different positions. The fish stayed too close to the fish in front to get the most benefit from the vortices they left behind (they should have stayed at least five tailbeats behind the pair in front) and not close enough to the fish on either side of them to get the most benefit from being able to push water against them (0.3–0.4 body lengths apart would have been the most effective).

These discrepancies in turn imply something about the selective pressures that have been operating on these fish. There might have been some selective advantage through energy saving during swimming, but it was almost certainly not as great as some other advantage – perhaps having a good view of approaching predators. Hydrodynamic advantages do not predict very accurately the structure of fish schools. Hence hydrodynamic advantages alone do not seem to have been the main selective force operating on the way fish position themselves with respect to one another. This particular adaptive hypothesis has therefore been discarded.

The basic assumption behind this fourth method of studying adaptation is that by discovering what the 'design features' of animals are or are not, we have, by implication, specified the reason why selection picked out the animals that are now with us rather than their rivals that once existed but which, lacking this key design feature, failed in the struggle for existence. The rival alternatives are

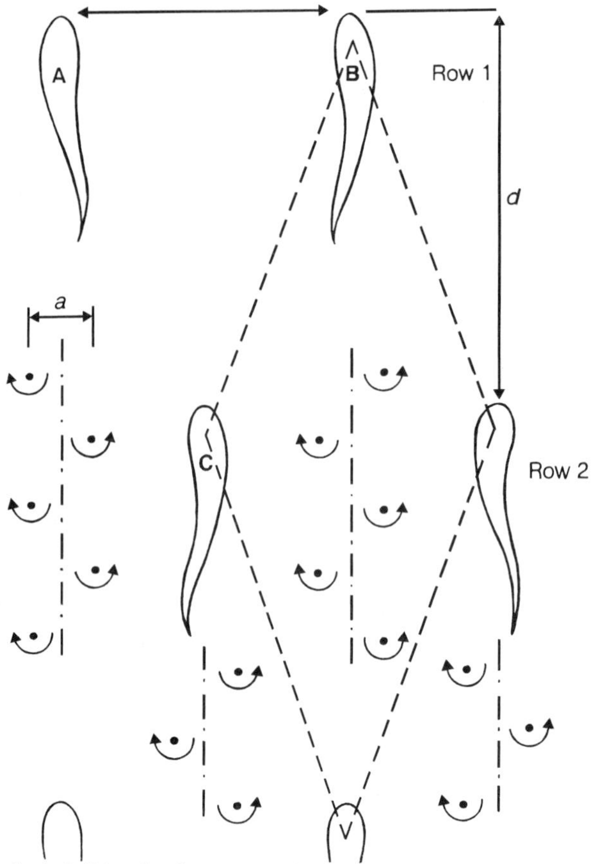

Fig. 1 Hypothetical fish school constructed on the assumption that fish position themselves so as to gain maximum energy saving from the swimming of other fish. Fish in the dotted diamond save energy in two ways: (1) Fish C, midway between A and B, receive induced velocity from the spinning vortices shed by A and B. These vortices do not become stable immediately, so fish C should stay more than five tail-beats behind the pair in front (D). (2) Fish swimming close together can push off one another, thereby reducing energy expended. Maximum saving occurs at a separation of 0.3 body length (*l*). In three species of real fish – saithe (*Pollachius virens*), herring (*Clupea harengus*), and cod (*Gadus morhua*) – fish kept two to three times too far away from lateral neighbours as expected from the hypothetical school and swam too close to pairs of fish in front of them. (From Weihs 1975, by permission of Plenum Publishing Corp.)

no longer with us in the flesh and we have to infer their existence from the intensity with which selection has apparently operated.

The more ruthless the selection has been, the more clear cut will be the design features and the more animals that did not possess them will have fallen by the wayside. Good design in the few that remain can only have been achieved at the cost of the many failures that were good at survival and reproduction, but not quite good enough.

By finding out how good the design is in the animals that are alive now, we get an idea of what selection has done in the past. We infer the existence of the failures of the past from the successes of the present. The ghosts of bats that could catch insects but not particularly successfully are resurrected by our understanding of where their shortcomings lay. And these ghosts, these failures, reveal the adaptations of the animals we now see around us. That is the idea behind using studies of modern animals to draw conclusions about the long-term actions of natural selection.

Conclusions: is all behaviour adaptive?

There are, then, four well-worked out and tested methods for studying adaptation. Two are what have been described as 'direct' because they involve trying to see death and destruction happening and re-create the causes of past reproductive failure again in the present. The nature of the adaptation is revealed when the rivals – real or artificial – are actually seen to die or fail to reproduce. We then know the reason why they died or failed to reproduce.

The other two, less direct methods concentrate on the living winners and try to infer from the characteristics that make them successful what it was that gave their ancestors the edge over long-dead rivals. One method compares different species with each other, searching for correlations with habitat and way of life. The other compares animals with a hypothetical alternative – a machine 'designed' to certain performance criteria. The closer a real animal conforms to the same criteria, the more confident we can be that natural selection, too, was favouring animals that had these characteristics and not others.

In other words, it is the search for the causes of death and reproductive failure that is the essence of the study of adaptation and that search is not an undisciplined quest for 'Just-So' stories as Gould & Lewontin (1979) implied. On the contrary, it is a way of looking for evidence that will either support or refute the specific hypotheses that are put forward to explain the evolution of any given trait.

Gould and Lewontin did, however, go further in their criticisms of adaptive hypotheses and in so doing have confused many people both inside and outside biology about the current status of the theory of natural selection and adaptive hypotheses in particular. For example, they claimed that there are serious alternatives to natural selection such as genetic drift. Genetic drift refers to the possibility that gene frequencies within a population may change not just because one genotype is adaptively superior to one another (natural selection) but because purely chance factors may lead one genotype to dominate the other. We can draw an analogy with tossing a coin. If we tossed a coin 10 times and it came down heads more than 5 times, we might be tempted to conclude that the coin was biased (that is, one side was the equivalent of being adaptively superior to the other). But even with a completely unbiased coin, it is unlikely that with every 10 tosses, there would be exactly 5 heads and 5 tails because sometimes, by chance, there might be deviations from the expected average. If we tossed the coin a very large number of times, we would expect that, on average, there would be an equal number of heads and tails, but with small runs we should not be surprised if a sequence of 10 tosses sometimes had 6, 7 or even 8 heads. The geneticist Sewell Wright argued that equivalent random effects occurred in real populations and that random drift of gene frequencies might lead to changes that had nothing to do with natural selection. While such effects may be very important in small populations (as they are in small numbers of coin tosses), their importance in larger populations is controversial. It is certainly true that where natural selection neither opposes nor favours a shift in gene frequency (as, for example, if a mutation affects part of a DNA molecule or a protein that has absolutely no effect on any phenotypic character), then drift may be important. Indeed, such 'hidden' mutations appear to be quite common (Kimura, 1991). But where a mutation has a clearly beneficial or clearly harmful effect on the organism, natural selection is likely to step in and oppose random changes.

In any case, genetic drift does not provide an explanation for complex, adaptive design in animal structure and behaviour. It is extremely unlikely that the detailed design of bat echo location calls, for instance, got there by random fixation of the many different genes that are involved in their production because at every stage of bat evolution, drift is as likely to make the calls worse for echo location as better. Only natural selection can explain the extreme complexity of their design (R. Dawkins, 1986).

Another of Gould and Lewontin's criticisms of adaptation that has been seized on particularly by non-biologists is that many traits are not adaptations at all, but 'spandrels' – that is, they are by-products of selection for something else. Spandrels are the triangular-shaped panels formed by the intersection of two rounded arches in churches or other buildings with domed roofs. They are a necessary by-product of mounting a dome on rounded arches and have no function in themselves. They are simply a way of filling in a space between two structures that are functional, namely, the arches holding up the roof. Gould and Lewontin argue, with some justification, that many structures in biology are similarly functionless by-products. The colour of blood is a spandrel, for instance. It makes no sense to ask what is the adaptiveness of blood being red because the colour is a by-product of selection on blood for something else, specifically, its oxygen-carrying capacity. However, blood itself is definitely not a spandrel. It has been heavily selected for and although its redness is an incidental feature of the iron that gives it its capacity to pick up molecules of oxygen, it would be misleading to use this as an argument against the adaptive nature of many features of blood. The possibility that there may be other biological spandrels does not in any way detract from the general conclusion that most traits are adaptive and can be shown to be so. There could not be by-products unless there were products as well.

Gould and Lewontin also did a service by pointing out that some biologists had been too ready to jump to adaptive conclusions without proper evidence. But that is quite different from the claim that adaptive hypotheses are not *in principle* testable. Despite rumours to the contrary, the theory of natural selection is today stronger and better supported by evidence than it has ever been (Dawkins, 1986; Endler, 1986; Williams, 1992). The hypothesis that something is an adaptation can be tested in the four ways we have discussed in this chapter. This hypothesis is not seriously threatened either by the idea that random genetic drift may sometimes occur or by the idea that some traits are by-products of others.

It may have struck you, however, that all the four methods for studying adaptation that we have discussed in this chapter refer to events that took place in the past. Modern animals were constantly referred to as being the descendants of animals that were once successful, as though their own evolutionary battles had been fought and won for them by their ancestors. But is this right? Surely, times – and environments – change? What was adaptive for an animal's ancestors in the past may now no longer be adaptive for its descendants in the present. This difficulty and some other apparent flies in the smooth ointment of adaptation will be discussed in the next chapter, once we have extended the idea of 'adaptation' to the even more misunderstood one of 'optimality'.

FURTHER READING

Tinbergen's (1963) characteristically clear-sighted paper gives an excellent introduction to ethology in general and to the study of adaptation in particular. Gould & Lewontin (1979) criticize some of the ways in which adaptation has been assumed rather than demonstrated and R. Dawkins (1982, 1986) discusses constraints in adaptive perfection. Harvey & Pagel (1991) advocate the Comparative Method and Hoogland & Sherman (1976) show how advantages and disadvantages can be studied in the field.

2

OPTIMALITY

Meerkats (*Suricata suricatta*) catch scorpions without being stung. Photograph by David Macdonald.

Growing out of the idea of adaptation is the widespread belief that animals may not just be 'well designed' but that they may even be 'optimal'. Looking at research papers published since the 1970s can give the impression that almost everything that animals do has at some time or another been described as 'optimal'. We find optimal foraging, optimal reproductive strategies, optimal times for threatening an opponent before retreating, and so on.

This emphasis on animals as 'optimizers' has led to an extraordinary degree of confusion about what 'optimal' really means and we cannot leave our study of adaptation until we have attempted to sort that out.

THE TWO MEANINGS OF OPTIMALITY

It is quite impossible to understand what 'optimal' means without realizing that it has not just one but two quite separate meanings.

Firstly, there is 'optimal' meaning 'leaving the most viable offspring in a lifetime'. This can be called the long-term meaning and it is concerned with the reproductive success of an animal over its entire life compared to its rivals, along the lines we were discussing in the last chapter.

Secondly, there is optimal in the short-term sense meaning that the animal appears to be optimizing some function in its day-to-day life, such as the amount of energy it is collecting in a certain amount of time. This is the sense in which the term is used in, say, 'optimal foraging'. The animal's ability to leave offspring may not be measured, but the animal is still described as 'optimal' if its behaviour conforms to some criterion, such as taking the most energy-efficient route to get to the most energy-rich food.

Of course, it is not strictly true to say that these two meanings are absolutely separate. Gathering food efficiently might well have a lot to do with leaving offspring – it could be the single most important factor contributing to it. But the connection is not straightforward. An animal that gathers food optimally might actually leave fewer offspring in its lifetime than an animal that gathers it less than optimally because it is so intent on feeding that it gets eaten by a predator. In other words, the long-term reproductive success kind of optimality and the short-term efficiency kind of optimality should be kept distinct. The thread that binds them, although it is undoubtedly there, may be a very tangled one indeed. Before we can unravel it, we have to look more closely at these two different senses in which 'optimal' can be used.

Long-term optimality

In the last chapter, we saw that our current picture of evolution is that, of the many mutants that have arisen in the past, the better ones prospered and it is their descendants that are with us today.

Now, for some people, this is equivalent to saying that the animals we see are the best of the ones that could ever exist, because given enough time, all possible mutations which could arise in a population will arise. Consequently, all possible mutations will have been given their chance and either failed in competition for some reason or demonstrated their success by leaving descendants which are still alive. The fact that a postulated mutant is not found in a population is, on this view, taken as evidence that it would not be successful because it has almost certainly arisen in the past, been given its chance, and failed.

This assumption is critically dependent on the further assumption that the environment has been stable for long enough to allow the slow processes of mutation and recombination to have come up with the best mutant the species is capable of producing. Hundreds and thousands of freaks and oddments, most of them less good at surviving and reproducing than normal members of the population, have to be born and try their luck. Only a few, a very few, will do better, and it may take a long time for these to appear.

There are, however, powerful reasons for suspecting that not many environments have been stable for long enough for that to have happened, long enough that is, for each possible mutation to have occurred at each site on the genome that could possibly affect the body of the animal. Certainly the broad general features of environments may well remain outwardly much the same for many generations. (Rain forests do not suddenly dry up, for instance.) But environments are not just physical things like temperature or rainfall. They also consist of other animals – competitors of the same species, predators and parasites out for a meal – each of them engaged in their own struggle for existence and each of them providing a constantly changing environment for the other animals in contact with them. Just as an animal has evolved an effective defence against a parasite, selection may then favour a new type of parasite that can get round the host's defences. Once this has happened, selection will then start favouring hosts that evolve new defences, and so on. Every time the optimum is approached, it slips away out of reach, with the parasite and its host engaged in a constant evolutionary dance, always on the move in evolutionary time. This idea has even been called the Red Queen hypothesis (van Valen, 1973; Bell, 1982) after the Red Queen in *Through the Looking Glass* who was always running hard just to stay in one place.

We might think, then, that we should define optimal in the long-term sense not as the best-possible mutant that could ever arise but more simply as the best of the mutants which have so far arisen in the present environment, keeping in mind the fact that the 'present environment' might not have been there for very long. This would leave open the question as to whether, with a bit more time, better animals than our present ones might arise. But this is not quite the same as the generally accepted meaning. More commonly, a statement that an animal is 'optimal' is taken to be a restatement of Darwin's theory of natural selection with a time element built into it. Optimality is a state which animals are said to approach when natural selection has been in operation for a certain length of time in the same environment. The possibility that parasites, predators, and rivals may see to it that the same environment rarely exists for very long is not taken as seriously detracting from the power of this formulation. It is assumed that the animals we now see are indeed optimal, subject to provisos such as these time-lag effects (Maynard Smith, 1978; Dawkins, 1982).

For the reasons we discussed in the last chapter, it is quite possible, although difficult, to study adaptation in the long term. Comparing the reproductive success of different kinds of animals may take years and generations, but at least in looking to the long term, we can be confident that we are dealing with the real stuff of evolution – with life and death, reproduction, and failure to reproduce. What, then, are we to make of optimality in the other, short-term sense? In experiments in which the reproductive success of an animal is not even measured, on what grounds can the animal be described as 'optimal'? The fact that large numbers of studies do conclude that animals are behaving optimally when those animals have not been allowed to breed at all must mean that 'optimal' is then being used in a different 'short-term' sense altogether.

Short-term optimality

To illustrate the meaning of short-term optimality, let us consider one early example of how an animal has been thought to be optimal. (More recent models have become both more realistic and more complex.) Gaining food energy is vital for survival and this classical version of Optimal Foraging Theory (or OFT) suggested that animals should optimize the net amount of energy they obtain in a certain time. For example, a bird taking food from a small tree and gradually eating up all the food that is there would eventually come to a point when it would gain more energy if it left that tree and flew to another, where the food had not been depleted, even though the flight would itself use up valuable energy.

According to OFT, the bird should balance the energy gained in the new tree against the energy lost through flight. If this is correct and such a bird really is optimizing its net rate of energy gain, then it should be possible to predict precisely when the bird should leave one tree and fly to another, provided that the crucial variables – rate of gaining energy in the old tree, energy cost of flight between trees, and expected rate of gaining energy in the new tree – are known. It is like trying to work out what determines the route of a particular travelling salesman who is required to visit a number of different towns. If we have a hypothesis that in working out which order to visit them his only consideration is to cover the smallest number of miles, we can predict which route he should take. If he does take this route, even though it takes him much longer than if he chose a more circuitous but traffic-free road and means that he misses the best of the scenery, then we can have some confidence that we have discovered the true basis on which he makes his decisions.

Cowie (1977) used this OFT model to predict how long great tits (*Parus major*) should spend foraging in a given 'patch' (a sawdust-filled cup in which pieces of mealworm were hidden; Fig. 2) depending on the quality of the 'patch' (the number of mealworms in the cup) and the 'travel time' between patches. Knowing how much energy the birds used in moving between patches and the energy they gained through the food they were able to find, he could predict reasonably accurately when they should leave one patch and go on to another. This strongly suggests that the birds were optimizing net rate of energy intake.

Fig. 2 Great tit (*Parus major*) foraging in experimental 'patch'. Photograph by Richard Cowie.

At this point, we must be careful not to become sidetracked by the issue of whether the detailed predictions of OFT are borne out in practice (Krebs & Kacelnik, 1991). Our main purpose is to answer a different question. If an animal optimizes energy intake in the short term, does this tell us anything about its being optimal in the long-term, evolutionary sense? If it can be shown to forage optimally, in other words, have we learnt anything about the way natural selection acted on its ancestors in the past? And, conversely, if it does not forage optimally, does this indicate that our ideas about the adaptiveness of the behaviour were wrong after all?

The link between short-term and long-term optimality

In order to understand how short-term and long-term optimality may be related, let us postulate a hypothetical animal. This animal (a great tit, say) has been shown to be an optimal forager under a wide range of circumstances. It has been tested with a range of foods differing in energy content and in the time needed to search for and deal with them. It has been placed in conditions with long, energy-consuming journeys between one patch of food and the next and in others where the trip is short and easily accomplished. In all cases the bird behaves in a way which gives it the greatest amount of energy that it is theoretically possible for it to gain within the constraints of its own anatomy and physiology. Short of evolving a new more energy-conserving way of flying or a different kind of beak that enabled it to eat its food more quickly, there is no way for it to do any better. It is in this sense an optimal forager. What could we conclude from such a remarkable bird?

In the previous chapter, we encountered the idea that adaptation may be studied through the 'design features' of an animal's behaviour or morphology. If an animal closely resembles a human machine which has been 'designed' to certain criteria, this can be used to give an indication as to how natural selection has acted on the animal and what failures have been eliminated in the past. Inferences about adaptation are drawn even when no comparison is made between the reproductive success of different sorts of animal. The example we used was that of bat echo location. Conclusions can be drawn about the adaptive significance of bat calls without any need to observe starvation in alternative sorts of bat. The 'design criteria' of the echo location system were enough.

Could the same argument be used for optimal foraging? Could it be said that the optimally foraging animal is like a machine 'designed' to optimize energy intake? And if it is, could we not make inferences about what natural selection has done in exactly the same way as we did for the bats? It is certainly true that very few optimal foraging studies look at reproductive success (but see Drent and Daan, 1980; Lemon, 1991). In fact, no one has shown that the lifetime production of offspring of an optimal forager is any greater than that of an animal which forages less than optimally. A comparable objection could be raised to the bats – that no measurements of reproductive success were made – and yet we were still able to draw conclusions about their adaptiveness. The direct methods for studying adaptation – comparing reproductive success of alternatives – would be just as difficult to apply to foragers of different degrees of competence as it would to bats with slightly different calls. But they may not even be necessary. Identifying the 'design features' of the behaviour might be sufficient to tell us what it is adapted to.

It is very important to look at the logic of this argument because it is one that is assumed to be valid by a very large number of people. The supposition is that there is a direct connection between the short-term and the long-term meanings of optimality. In other words, if we can discover what an animal is optimizing in the short term, it is assumed we have a direct insight into why natural selection favoured its ancestors in the long term, and penalized the rivals of those ancestors.

As we saw in the first chapter, the aim of studying adaptation is to discover why some animals lived and others died and short-term optimality studies are thought to do just that. The unsuccessful animals might have failed because they could not get enough food or because they were eaten by predators or because they made poorly constructed nests that could not protect their eggs. There are many different possible reasons to account for the success of some and the failure of others. The goal of the study of adaptation is to find out what those reasons were – to elucidate the causes of death and reproductive failure.

The linking of short-term and long-term optimality assumes that those causes of failure long ago remain permanently impressed on the modern descendants of those who were not failures. Animals that were engaged in a struggle for existence which depended primarily, say, on getting enough food (in other words, their rivals were dying or failing to reproduce from starvation rather than anything else), would be favoured if, in their own lifetimes, they concentrated their efforts on getting as much food energy as they could. Those that were more efficient at gathering food than others would do better and the mutant that was best at food collecting would do best of all. This highly efficient food gatherer might not be very good at getting away from predators, but if predation were a minor risk compared to starvation, this would not matter. If starvation were the main arbiter between life and death, then the efficient food gatherer would flourish and leave many descendants. One that optimized energy intake would leave most of all.

We now look at those descendants in our own time. We observe, let us imagine, that their entire behaviour appears to be 'designed' to maximize energe intake under all circumstances. Not having been present during the battle of their ancestors, we nevertheless infer that food was what the ancestral battle was about. The scars of that battle are still visible in their descendants.

That, at least in its simplest form, is the thread that binds the two sorts of optimality. The discovery of a 'design feature' or a function which is optimized in the short term is thus used as a way of re-creating evolutionary events that happened a long time in the past.

THE SEPARATION OF SHORT-TERM AND LONG-TERM OPTIMALITY

But what happens if the modern descendant does *not* forage optimally? Suppose, when we study its behaviour closely, we find that it is moderately good at collecting food but not really optimal. It wastes a lot of energy flying up into trees where there is no food. Can we conclude anything about its ancestors under these circumstances?

One obvious inference would be that getting enough food was not the only thing that made the difference betwen the success of its ancestors and the failure of its rivals. Perhaps going up into trees to look for predators also contributed, and the rivals that thought only of food (optimal foragers though they might have been) failed to leave as many offspring as the more wary animals.

There is a crucial point to be made here. The less-than-optimal forager that is aware of the approach of danger may, in the end, leave more offspring than the optimal forager whose eyes are never raised from its food. The optimal forager (optimality in the short-term sense) may therefore be less than optimal in the long-term sense because long-term optimality may demand a compromise between several different short-term demands. This is not a contradiction. It forcibly illustrates the fact that short-term and long-term optimality are logically quite distinct from one another and should not be confused.

Because they are logically distinct, we should be careful not to put too much reliance on short-term optimality theories that stress just one factor (such as optimal foraging theory) to tell us about long-term optimization. To achieve long-term optimality, animals may have to trade off optimal food finding against, say, optimizing their chances of getting a mate. The result may be that neither their search for food nor their gaining a mate may be done in the optimal way they could achieve if they could devote themselves single-mindedly to one activity or the other. The two activities may conflict, perhaps because they have to be carried out in different parts of the environment. Going to the place that is best for foraging may reduce the chances of finding a mate practically to zero while spending too much time looking for a mate may mean that there is no time left for feeding and starvation becomese a real possibility. The animal that achieves the optimal balance between feeding and mating will be the one that contributes most to the next generation and will therefore be optimal in the long-term sense.

If we did happen to find a real animal that took no notice of anything except acquiring energy optimally under all circumstances, then the conclusion that feeding efficiency was the only selective force that had been at work might be quite reasonable. Here short-term optimality (based on foraging efficiency) and long-term optimality (based on the number of offspring produced in a lifetime) would be much the same thing. In practice this is unlikely to happen, however, because in most environments, getting enough food is not the only criterion of success. Death comes in other forms besides starvation. Animals which are, on the average, best at evading all of these will become the progenitors of future generations. Their behaviour will be a compromise between getting enough to eat, avoiding being eaten themselves, mating, and so on (e.g. Ydenberg & Houston, 1986; Abraham, 1993).

So an animal may be 'optimal' in the long-term sense because selection will favour those that reach the 'best' compromise, but taken individually, none of its constituent behaviours may be optimal at all in the short term. Efficient food gathering may be compromised by the need to keep an eye out for predators. The best method for avoiding predators may be made impossible by the risks of dehydration or starvation they involve. The behaviour of the animal may, in fact, be an accurate reflection of the opposing selective forces that have acted on its ancestors in the past. But we should not be surprised if the picture we gain of what those selective forces were is not at all clear cut. It is only in the unlikely event of there being a single selective factor that caused death in some and favoured others that we would expect animals to optimize a single, simple function in the short term.

It is, therefore, a very difficult exercise, following the thread running from the behaviour of our present-day animals back into time to the selective forces which shaped their ancestors. As we have seen, it is not impossible to follow it, tenuous and tangled though it often is. The way animals are 'designed' can be used to study adaptation but only if considerable care is exercised in interpreting the battles of the past from the survivors of the present. More recent optimality models do take into account the fact that success in reproduction will be dependent on more than just one short-term factor (Harvey, 1985; Houston & McNamara, 1990; Krebs & Kacelnik, 1991). The need for efficient foraging is acknowledged to be balanced against the need to look out for predators, the need to find a mate, and so on. But, of course, if the (long-term) optimal animal has to make compromises and, say, forage less than optimally because of the risk of predation, then it may arrive at the optimal compromise between feeding and predation, but neither feeding nor predation will (in the short-term sense) be optimal. The optimality of the individual components (feeding, mating, etc.) may have to be sacrificed in the interests of survival and reproduction of the whole animal.

This, of course, raises the question of why anybody ever thought that any animal should forage optimally, since to think that they should implies that only one selection pressure – getting enough to eat – was the only one that mattered. Perhaps under some circumstances, such when food is scarce or when survival of the offspring depends critically on their being well-fed, it is. But usually, other selection pressures will creep in and reproduction will not depend just on getting that last little bit of food but on other things as well. The value of a specific optimality model such as Optimal Foraging Theory then becomes not so much in it being the last word in explaining behaviour but as a working hypothesis to see how far one factor does explain what the animal does. Suppose an animal did forage optimally except when predators are around. This would at least suggest that the selection pressure to feed efficiently was a very powerful one, only surpassed under some circumstances by the selection pressure to avoid being eaten. The adaptive nature of the animal's behaviour would have been revealed by starting with an over-simplified model, seeing how far it could get and then modifying it when its limits were reached.

It could be argued that this is a more fruitful way of studying adaptation than to insist right from the outset that all behaviour will be a web of compromises between different selection pressures and that there is no point in going any further. At least single-factor optimality models and their modifications provide testable predictions about the action of specific selection pressures even though their predictions may not be upheld in practice. Some progress is made even if they turn out to be wrong, whereas with the vaguer web-of-compromises approach, nothing is learnt at all. Of course it is unlikely that only one selection pressure operates on an animal, but as the goal of all studies of adaptation is to identify which (or which combination) of selection pressures has been operating and not merely to say that 'quite a lot' of them have, optimality modelling does provide a way forward. Being specific though perhaps wrong is usually more constructive in science than being so general that no one can tell whether you are right or not.

Nevertheless, the current use of the single word 'optimal' to cover both long-term and specific short-term processes is extremely confusing. It is essential to recognize the difference between short-term optimality and long-term optimality, for they are logically separate concepts. Only when the distinction between them is fully appreciated can the tangled thread be followed with any certainty.

'Rules of thumb'

Even with animals that appear to be behaving optimally in the long-term sense, there is another important reason why we should make a careful separation between long- and short-term optimality. This is that their optimal (long-term) behaviour may be achieved by their following relatively simple 'rules of thumb' in the short term. For example, if a bird appears to be foraging optimally, then we might be tempted to think that it had a detailed knowledge of the amount of food that could be obtained from every single one of a large number of different trees, how much energy it uses in flying various distances and so on. But in fact the bird may achieve its optimal behaviour by a series of simple rules such as 'if no food found within 30 seconds, move to next tree'.

In some environments, this simple rule of thumb might be just as effective at giving optimal behaviour as a very much more complex set of rules or more detailed knowledge of the various parameters involved. If, say, the bird lived in a wood in which, on average, one in every five trees had five times as much food as the others, then a bird that left a given tree after a short time of finding no food would rapidly find itself in a food-rich tree even if it had no knowledge of the fact that one in every five trees was a great deal more profitable than the rest. The simple rule to leave after 30 unproductive seconds might, in that particular environment, lead to just as 'optimal' (short- and long-term) behaviour as a set of more complex rules. The key phrase here is 'in that particular environment' because in another environment, in which, say, none of the trees had much food, the rule 'leave after 30 seconds' would mean that the bird hardly ever found food at all and was forever flying onwards.

Rules of thumb are fine in the environment in which they were evolved but may lead animals to behave 'stupidly' when conditions change. The oystercatcher described by Tinbergen (1951) that ignored its own egg and tried to incubate an ostrich egg instead was probably following a rule of thumb such as 'sit on the biggest egg available', a rule that would work well in the oystercatcher's normal environment but which would break down in the artificial situation created by an ethologist placing an ostrich egg near its nest.

So one reason why animals may sometimes appear to be behaving less than optimally is that they are following rules of thumb appropriate to one environment but are being studied in a completely different one. Optimality in the long-term sense can only be studied in the environment in which the behaviour evolved, even though we may need laboratory studies to tease apart the precise short-term rules that give rise to it.

COSTS, BENEFITS AND ECONOMICS

There are some rather more elegant ways of expressing the ideas we have been discussing. Many biologists sound a bit like economists. They use words like 'costs' and 'benefits' and even 'prices' and 'budget constraints'. Their papers have titles like 'The coal tit as a careful shopper' (Tullock, 1971). They see an animal spending time on one activity or another much like a human consumer spending money on one commodity or another. Struck by the fact that economists assume that the organisms they are studying (in their case human consumers) are also optimizers, biologists have begun to phrase their ideas about evolution in economic terms. We will conclude this chapter with a brief look at some of them, but all we are doing is rephrasing and clarifying the ideas we have just discussed.

For economists, the function which is being optimized is known as 'utility', which is roughly the equivalent of 'satisfaction' or 'pleasure'. Human beings spend their limited incomes on the various options they can buy, and they are assumed to make choices (between buying cigarettes, say, or buying an exotic kind of food) in a way which gives them the greatest amount of this utility or satisfaction.

Biologists, enthusiastically embracing the economic terminology, directly translate the idea of utility into what they see as its biological equivalent. And that is where the trouble starts, for, as we have seen, there are the two different ways that animals have been thought to optimize. Sometimes the biological equivalent of 'utility' seems to be taken as long-term reproductive success, with animals being said to optimize this and to pay 'cost' and 'benefits' in terms of loss and gain of viable children (e.g. Rubenstein, 1980). At other times, it is taken as something more short term, like net energy intake per unit time. Here 'costs' means energy expenditure and benefit means energy intake. We have already seen that these different senses of optimizing have no simple relationship to one another. An energy 'cost' may well turn out to result in a 'cost' to offspring but not necessarily in any simple or one-to-one way. In other words, the animal which forages less efficiently may have fewer offspring, but, it may equally turn out to have more than the animal that forages more efficiently, through being warier about predators.

When applying the economic analogy, therefore, we must constantly keep in mind the distinctions of the previous section. McFarland & Houston (1981) have proposed some terminology to keep us on relatively firm ground. They suggest that we should take as the biological equivalent of utility (satisfaction) only the short-term optimality rules described by what they call the 'goal function'. 'Energy gained per unit time' would be an example of a goal function.

The longer-term reproductive success kind of optimizing, on the other hand, they suggest should be described as following a 'cost function'. If variation in the ability to get enough food is the only thing that contributes to the variance in reproductive success (the unlikely situation we imagined before), that would be equivalent to saying that unsuccessful animals died of not getting enough to eat and nothing else. The cost function would be an expression of this. A perfectly adapted animal in such circumstances would be one in which the goal function (short-term optimizing process) accurately reflected in the animal's day-to-day life this long-term need to get enough food; in other words, one in which the goal function, too, was geared exclusively to food finding. Cost function (long term) and goal function (short term) are thus assumed by McFarland and Houston to be in close accord in animals in their natural environment, although separable in principle as we have already seen.

Many people are put off by the mathematics involved. But the distinction made between different sorts of optimizing is exactly the one we have been groping for using the rather inelegant terms of 'optimizing in the long-term sense' or 'optimizing in the short-term sense'. The problem of the relationship between the two can be re-expressed as that between goal function and cost function.

We should not think, however, that just because we have found some mathematics to say what we wanted to say all along or that the ideas of optimality are also used by economists, that all the problems with it have disappeared. Optimality is still a difficult concept. Making the link between goal functions and cost functions is just as hazardous as trying to connect short-term with long-term optimality. Fancy economic terminology should not make us lose sight of the genuine problems that are involved.

Conclusions: do animals behave optimally?

It should by now be clear that the answer to this question will depend on which sense of 'optimal' is being used. In the short-term sense, we would not expect animals to forage optimally, look out for predators optimally, mate optimally simply because doing one of these in an optimal way will almost certainly interfere with their ability to do the others optimally too. They will be jacks of all trades and optimal in none. Or rather, only in the unusual circumstances that one selection pressure is so strong that it dominates their lives and chances of reproduction would we expect the perfect alignment of cost function and one goal function that would lead to short-term optimality.

In the long-term sense, too, we should not be surprised to find animals behaving less than optimally. They may achieve a less than optimal compromise between their different component behaviours for all sorts of reasons, an important one being the 'time lags' we mentioned in the last chapter and discussed in this one in connection with the Red Queen hypothesis. If an environment is changing rapidly, animals may be well adapted to an environment of the past but poorly adapted to that of the present. In other words, they may be optimal (in the long-term sense) to an environment which is no longer here and genetically have simply failed to keep up. The result will either be extinction or progressive change over the generations. How important you think these time lags are in 'spoiling' optimality will depend on whether you take the 'Red Queen' view of evolution as a constant and desperate race to stay in one place, or whether you see animals as generally adapted to their present environments, with just a few little blips, such as inability to cope with a specific new disease or the advent of motor cars. And even if they are well adapted to their present environments, many animals follow 'rules of thumb' that make them appear suboptimal if their behaviour is observed in unusual conditions.

There are other reasons, too, why animals may appear to us as not behaving optimally even in the long-term sense. We saw that one way of defining optimality was to say that it was the best of the mutations that have so far turned up in a given environment. So one reason for an apparent lack of optimality or adaptiveness may be that the relevant mutation has not yet turned up. Perhaps it never will because it would be physiologically impossible or demand so much rearrangement of the animal's body that animals with that mutation would be at a disadvantage for other reasons. Historical constraints of this sort may be commoner than we realize, particularly with traits that demand changes in many different genes. Human beings were able to move from propeller-driven aircraft to jet planes by literally going back to the drawing board and starting again. But if they had been forced to do the equivalent of evolving from one to the other, they would have had to make intermediate machines that differed in small ways from the original propeller plane, changing one bit at a time until they had built a jet aeroplane, with the constraint that every single modification not only had to fly but to fly better than the version before! (Jacob, 1977). Animals may therefore be left with the equivalent of propellers when jet propulsion would be optimal simply because they are stuck on one peak of what Sewell Wright (1932) called an adaptive landscape. Another peak may be higher (more adaptive), but if getting there involves going down into a valley (being temporarily less well adapted than competitors), it may never be attained. There will be selection against any who try.

FURTHER READING

Parker & Maynard Smith (1990) and Krebs & Kacelnik (1991) show what optimality studies try to achieve. McNeill Alexander (1982) shows how optimality ideas can be applied to a broad range of structure and behaviour. Ydenberg & Houston (1986) discuss optimal trade-offs and R. Dawkins (1982) lists some constraints on perfection.

3

INCLUSIVE FITNESS

African elephants (*Loxodonta africana*). Photograph by Matthew Evans.

Throughout the last two chapters, we have looked at natural selection in terms of the survival of the individual and the number of surviving offspring it produces. We have stressed that what is important is an animal's lifetime production of offspring that themselves survive to reproductive age. This is one measure of 'fitness' but it may have given rise to the feeling that the picture is incomplete. We have made no mention of relatives besides offspring such as siblings, nieces, or nephews. In view of the importance attached to behaviour which benefits these other sorts of kin in much of the recent literature on behavioural ecology and animal behaviour, we surely should have done. We should, in other words, have mentioned not just fitness in terms of offspring but taken into account other sorts of relatives and included a measure of what Hamilton (1964) called 'inclusive fitness'.

Yes and no. Yes, aiding relatives other than offspring may be important. But no, this does not mean that counting only direct offspring is an invalid measure of evolutionary success. They amount to much the same thing.

This chapter will be about justifying this apparently paradoxical statement and showing that counting the number of offspring was in fact a central part of what Hamilton's much cited paper was all about. We will further see that it was also about shifting the emphasis from selection at the level of the individual to selection at the level of the gene.

AIDING RELATIVES AND 'INCLUSIVE FITNESS'

Hamilton (1964) used the idea of 'inclusive fitness' as a way of calculating the conditions under which a gene might spread through a population, taking into account the effect that bearers of that gene might have on different sorts of relatives. In a moment we will see exactly how Hamilton defined 'inclusive fitness', which is what has been so greatly misunderstood. The misunderstandings that have arisen are not minor errors that can be quickly deleted from a few papers. They are (not to put too fine a point on it) major blunders. In 1982 Grafen listed fourteen commonly used textbooks on animal behaviour that had either defined inclusive fitness wrongly or failed to define it adequately. A quite extraordinary state of affairs! A central idea about animal behaviour misused and apparently misunderstood by almost everybody who discussed it. We are going to have to do quite a lot of unravelling. What follows is largely based on Grafen's (1982, 1984, 1991) papers and is based on the assumption that what experienced biologists misunderstood then could well be misunderstood now by students coming to it for the first time.

We start, where we should, with Hamilton (1964). Hamilton pointed out that there is nothing special, from a genetical point of view, about children. A child has a 50 per cent probability of sharing a given rare gene in common with its parent but then so it does with its siblings. More distant relatives also have a fair chance of sharing this gene, the chance being directly calculable from a knowledge of how closely related they are. This chance of sharing a rare gene is called the 'coefficient of relatedness', which is called r.

Hamilton was not, of course, the first to realize this. But he was the first to provide a way of calculating whether a genetic tendency to help relatives would be successful enough to spread through a population. His purpose, then, was very similar to that of an 'ordinary' geneticist trying to calculate whether a gene for, say, white eyes in a fruitfly, would spread through a population of red-eyed flies. Geneticists talk about the fitness of a gene as its success relative to that of an alternative, in this case, the ratio of white-eyed genes to the genes giving rise to the normal wild-type red eyes.

Mutant genes for white eyes, however, have effects which extend no further than the body of the fly actually bearing them and which affect only that body's chance of leaving offspring. In the case of 'relative-helping' genes, on the other hand, the effect is to help copies of themselves (and other genes) inside other bodies. These genes help the other bodies to have more offspring, which in turn will have a high chance of carrying the relative-helping genes.

The calculation of whether such genes will spread is therefore quite complicated and has to take into account these 'other body' effects as well as the effects the genes have on the bodies that actually carry them. Hamilton coined the term 'inclusive fitness' to cover both these 'own body' and 'other body' effects. He argued that the chances of a relative-helping gene spreading through the population could be expressed by the equation

$$rB - C > 0$$

where B is the enhancement to the relative's chances of surviving and reproducing as a result of help being given to a relative and C is the cost to self as a result of doing the helping. Relative-helping evolves when the gain in fitness to the relative (devalued by degree of relatedness to the helper) outweighs the loss in fitness to the one doing the helping. The inclusive fitness of the helper takes into account both of these factors.

As will become clearer as this chapter goes on, inclusive fitness is in one sense a measure of gene frequency (gene success relative to some standard) but it is expressed in terms of a property of individual animals. What is really going on is selection at the gene level, but many people want to continue to talk about selection at the individual level and the 'inclusive fitness' of individuals. However, although this is intuitively reasonable, it is also the major source of confusion about inclusive fitness.

Inclusive fitness, seen at the gene level, is just 'ordinary' geneticists' fitness with room for the effects on the reproductive success of relatives included but put in such a way that it can be applied to whole animals. If this is to be done correctly, we have to be sure that fitness calculated at the gene level and inclusive fitness calculated at the individual level give the same answer. Doing this means engaging in the intellectual contortions that have led to such major misunderstandings. Hamilton (1964) described the situation in this way:

Inclusive fitness may be imagined as the personal fitness which

> an individual actually expresses in its production of adult offspring as it becomes after it has been first stripped and then augmented in a certain way. It is stripped of all components which can be considered as due to the individual's social environment, leaving the fitness which he would express if not exposed to any of the harms or benefits of that environment. This quantity is then augmented by certain fractions of the quantities of harms and benefits which the individual himself causes to the fitnesses of his neighbours. The fractions in question are simply the coefficients of relationship appropriate to the neighbours whom he affects: unity for clonal individuals, one-half for sibs, one-quarter for half-sibs, one-eighth for cousins, . . . and finally zero for all neighbours whose relationship can be considered negligibly small.

The key to understanding what Hamilton meant, and why people have misunderstood the connection he was making between genes and individuals, lies in the curious business of 'stripping' and 'augmenting'. First, he says, inclusive fitness has to be 'stripped of all components which can be considered as due to the individual's social environment'. This means that we have somehow to work out how many children an animal would have had if he had no relatives around either to help him have children, or to hinder his rearing of children. Imagine, as an illustration, an animal that is helped by its brother to have children. The two brothers set up territories next door to each other. One brother rears some children of its own, but spends a lot of its time going next door to feed its nephews and nieces. This is time which it might have spent feeding its own children. It thus has fewer offspring of its own than it might have done, whereas the brother who is helped has more.

We have to try and work out the gains and losses resulting from this behaviour. According to Hamilton, we have to start by working out how many offspring an animal would have had by its own unaided efforts if it had been neither harmed nor helped by anyone else. We have to 'strip away' all the other offspring. We must ignore the ones that owe their existence to help from relatives and we must try to count the unborn ones that would have been alive but for hindrance from relatives.

The first part of this stripping is relatively straightforward. The offspring that were gained by the helped brother must be stripped away from him, because they would not be alive but for help from their uncle. We might even be able to get an estimate of how many of these extra offspring there were by comparing the reproductive success of nests with and without brothers that helped (Emlen & Wrege, 1989).

The second part of the stripping operation is more difficult both conceptually and in practice. We want to know how many offspring the helper brother lost as the result of the effects from his social environment. This does not mean the number that he lost as the result of his own efforts at rearing nieces and nephews, but the number he lost *as the result of the actions of other animals.* For example, if his brother inveigled him next door and actually prevented him from having young of his own then this would be a 'harm from the social environment' and the offspring lost as a result of this should be stripped away. Stripping away unborn offspring means (through the double negative of removing something which has been taken away) effectively giving them back to the helper brother because we want to know how many offspring he would have had but for the interference from the helped brother.

The offspring that the helping brother lost as the result of his own actions in tending nieces and nephews are not part of this stripping operation because they are not casualties of the social environment. If there were no inveigling by the helped brother, the helper brother could not be said to have been harmed (he did not lose offspring) by the genotype of the brother to which he gave help, but he was harmed by his own genotype, which led him to give so much of his attention to his brother's children that he lost out on his own. We are stripping only those offspring that are affected – for better or worse – by variation in the genotypes of *other individuals* and trying to leave ourselves with an estimate of offspring that are affected only by variations in the genotype of the individual in question. We are concerned with offspring produced alone, unaided, and unhindered.

Secondly, we come to what Hamilton called 'augmenting'. To the offspring produced alone and unaided, we have to add relatives (devalued by the coefficient of relatedness) that are around because of the effects of relative-helping behaviour. Notice that this does not mean all relatives that exist. A solitary individual rearing offspring on its own without aiding its brother in any way is probably going to have nieces and nephews just because its brother (on the other side of the wood) is probably going to reproduce. Just having nieces and nephews is no real achievement. What matters, as far as the spread of brother-helping genes is concerned, are the *extra* ones an animal has as the result of helping. These are what could potentially give the evolutionary edge over the solitary kind of animal.

So we have to know how many extra offspring there are. In other words, we want the helped brother's 'stripped' offspring total first and then the augmenting effects of his helpful brother. Then we need the helper brother's stripped offspring total and have to 'augment' that, negatively, by the number of offspring he would have had if he had not been rearing nieces and nephews.

The purpose of the stripping and augmenting is to make absolutely certain that we have correctly tracked the changes in gene frequency. The big danger is to count the same individual twice and so to get an inflated idea of the frequency of the genes that it is carrying. For example, a single offspring of a brother which is helped could, unless we are very careful, be counted once as the son of his father and then again as the nephew of the uncle who reared him. It would look then as though there were two individuals where there is really only one and the result of 'brother-helping' would appear to have twice the genetic benefits that it really does. If a nephew owes his existence to the efforts of his uncle, then 'stripping' that nephew away from his real father makes quite certain that he does not get counted twice.

Unfortunately, there are many erroneous definitions of inclusive fitness that do inadvertently go in for double-counting and so give an inflated idea of the benefits of helping relatives. For example, in the 1981 edition of their otherwise admirable textbook, (since, I am glad to say, revised), Krebs and Davies put in their definition of inclusive fitness that it 'includes the animal plus 0.5 times its number of brothers and sisters plus 0.125 times its number of cousins and so on'. In other words, they are recommending the inclusion of all non-offspring relatives (suitably devalued by *r*, the degree of relatedness), not just the extra ones that owe their existence to help from relatives, as well as the inclusion of all offspring, not just the ones that the animal has by its own efforts. Grafen (1984) refers to this erroneous definition as the simple weighted sum as it involves adding up all the relatives and devaluing them by their degree of relatedness.

If this definition is used to calculate the inclusive fitness of all animals that help relatives, then it will give the wrong answer. It will count many individuals twice or even many times over, as the offspring of the parent who had them and as kin of their relatives. This will give an exaggerated idea of the advantage of relative-helping, falsely inflating the sum of the relative-helping genes around.

If we keep our eye on the original, geneticists' idea of fitness as the number of copies of a gene in the population relative to the number of copies of its allele, we can see how very misleading this would be. We must not count a single copy of a gene twice just because it occurs in a single individual that could either be described as someone's son or as someone else's nephew. The son and the nephew are one and the same individual. He does not suddenly sprout more copies of his genes because he could also be described as the great-nephew of a third individual, the grandson of a fourth, and the cousin of a fifth and a sixth. In order to preserve the correspondence between gene frequency (gene-level way of calculating evolutionary success) and inclusive fitness (individual-level way of calculating the same thing), we must resist the temptation for all this multiple counting.

If an individual survives to adulthood, not all his relatives can take credit for him – he must not be included in all their inclusive fitnesses because when these are summed together to give the inclusive fitness of the relative-helping *genotype,* this will be wildly inflated. The only way to avoid this error is to ask firmly by whose efforts the individual managed to arrive at his adult state. If his parent, by his own effort, reared and cared for him, then he can be counted as part of his parent's inclusive fitness and should not be included in that of an uncle nesting many miles away who has had no effect on him. If he would have died but for the care of an uncle, however, he should be counted towards the uncle's total, not the parent's. Then we must not forget that the uncle, in order to rear this nephew, probably had to forgo at least some reproduction of his own and may even have suffered hindrance from his relatives. He may have gained some kin, but he has had to pay for them in the hardest currency of all – lost offspring of his own.

Offspring and inclusive fitness

Grafen (1982) documented a number of papers where inclusive fitness has been wrongly calculated. The commonest fault is to use a version of the simple weighted sum although, until recently, the actual measurement of this in the field has been so difficult that not too many man-hours were lost through the pursuit of it. Only rarely is it known which animals are the cousins, nieces, nephews of which other ones.

But, if the simple weighted sum is difficult to measure, proper inclusive fitness in the sense that Hamilton defined it appears to be quite impossible. We seem to have to know not just what does happen but what would have happened under various unlikely circumstances such as if a socially living animal were suddenly stripped of all the harmful or beneficial effects of its social environment. We appear to have to measure not just how many offspring, nieces, nephews, and so on an animal has (difficult enough in itself) but actually how much it has contributed to the fact that they are alive and well and available to be counted.

Fortunately, there is an easy way out of this difficulty. Counting the number of offspring gives approximately the same answer. A bland statement to this effect appeared at the beginning of the chapter. It is now time to justify it and to see why something as down to earth and easy to measure as the number of children should give us the same answer as all those flights of fancy about stripping and augmenting. The link is as follows.

In a group of animals, all of whom carry the trait for relative-helping, there may be a number of offspring that owe their continuing survival wholly or partly to aid given by relatives other than their own parents. In an individual case, it is often very difficult, as we have seen, to know exactly how many offspring a parent would have had without help or hindrance from relatives, and how many children the relative would have had itself if it had not been helping or interfered with by the animals around it.

Now, 'inclusive fitness', like ordinary fitness, refers not to the reproductive output of a single individual, but to the success or otherwise of a whole genotype compared to some alternative genotype. An individual's success or failure, including the effect that he may have on relatives, is therefore important only in so far as it contributes to the average reproductive success of that genotype. And 'reproductive success' means adult offspring produced. By definition, the fitness of a genotype is increased relative to that of another if the number of copies of that genotype passing into the next generation is greater, and the parent–offspring line is the only way in which this passage can be made. Genotypes would go extinct if none of their representatives produced offspring. So we are not talking about a genotype that only helps relatives and never reproduces itself. Rather, we are talking about a genotype that may reproduce itself but can switch to aiding relatives in those circumstances (to be discussed below) when such aid would lead to a higher representation of that genotype in the next generation than by reproducing itself. We are, in other words, talking about a gene that sometimes makes the body it is in produce offspring and sometimes that very same gene will be in the body of an animal that is sterile but helps the reproduction of other bodies that have a high chance of containing the same gene.

What the gene makes its body do (reproduce or help) will be determined by environmental factors such as food or pheromones. Sometimes the switch occurs early in life and is irrevocable (as in the formation of sterile castes in the social insects). Sometimes the potential to do either remains throughout life (as with naked mole rats (Jarvis, 1981)).

What is of interest, however, is whether and under what circumstances the genotype for sometimes helping relatives to have offspring at a cost to the self's offspring results in a greater total number of offspring for that genotype than a genotype that never helps relatives. If we can understand this, then we have the key to understanding the whole of kin selection and inclusive fitness.

CHESTER COLLEGE LIBRARY

The trick is to think not in terms of individuals but of all the individuals making up a given genotype. The *average* inclusive fitness of all the relative-helpers is given simply by the number of adult offspring they collectively produce compared with animals that do not aid relatives. The genotype as a whole does not aid non-offspring relatives as a substitute for rearing offspring. It aids relatives so as to produce more offspring. Some members of the genotype will have fewer offspring as a result of helping other individuals. But if the helped individuals have more than enough extra offspring in consequence to make up for this loss, the genotype will prosper. Its average fitness will be greater than the alternative if the average number of adult offspring produced by the genotype as a whole is higher than that produced by an alternative genotype in which no individual aids relatives other than its own offspring.

The number of adult offspring produced by a genotype will automatically reflect the extent to which relatives have helped as well as the extent to which their own reproduction has suffered. It will effectively have already been stripped and then augmented again by all the elusive things we find it so difficult to discover for a particular parent or a particular uncle. It will be the result of the net gain in adult offspring summed over all the individuals of the relative-helping type. As much help will have been received from relatives as will have been given to relatives, on the average for the genotype as a whole.

Consequently, the stripping we have to do for the average relative-helper (calculating how many own offspring would exist in the absence of help or hindrance from other relatives) and the augmenting we have to do (adding in the harm or benefit done to relatives) exactly balance out when we do the sums for the whole genotype, leaving us with the number of adult offspring actually produced.

The importance of concentrating on average inclusive fitness of the relative-helping genotype, rather than one individual's own inclusive fitness, can be seen by considering an extreme case of relative-helping – the sterile castes of ants, bees, and wasps. A sterile social insect worker has no offspring – in this sense no fitness at all. But its genes are passed on. The gene that appears in sterile workers and makes them aid their sisters without reproducing themselves is the same gene that, in the bodies of young queens, gives rise to highly fertile individuals. Whether a bearer of the gene is sterile or fertile will be determined by the environment (food, pheromones, etc.) experienced by the individual in its lifetime. The inclusive fitness of the whole genotype can be very great indeed because many fertile offspring are produced as a result of this arrangement.

To work out the inclusive fitness of such a genotype, we need to know how many fertile adult offspring are produced, on average, by bearers of this conditional (i.e. switchable between fertile and sterile) gene and compare that with the average number produced by an alternative, such as bearers of a gene that always gives rise to fertile adults, never workers.

Sometimes more fertile children can be produced by having some sterile children who are helpers than by making all children into fertile ones. The actual number produced is despite loss of reproduction by the sterile workers and because of aid given by them. We do not have to go through the stripping and augmenting calculations because the net result of all the attendant costs and benefits will be reflected in reproductive output. If we count the number of fertile offspring of the two types (those with and without sterile offspring), this will tell us whether the average inclusive fitness of the one is greater or smaller than that of the other.

It should perhaps be pointed out that although comparing the number of offspring in this way does give a valid indication of the comparative evolutionary success of two genotypes, the average number of offspring does not exactly equal the average inclusive fitness to be found in Hamilton's equations. The reasons for this are technical and need not concern us here (Grafen, 1984), and they do not invalidate the idea of comparing offspring numbers rather than inclusive fitnesses. There may also sometimes be problems with counting offspring if it is not possible to identify which of them carry which genes (Grafen, 1982). It is then impossible to work out whether a given genotype is at an advantage or not simply by counting offspring. Provided it is possible to identify genotypes, however, this problem does not arise. The number of offspring is what really matters.

All this is, of course, most extraordinarily convenient. It is far easier, in practice, to count numbers of offspring than it is to count all the relatives that are demanded by the simple weighted sum. Yet a comparison of offspring numbers is a much better measure of evolutionary success as well as being closer to Hamilton's original idea 'of inclusive fitness'.

It might be thought, though, that what we have in fact done is to show that inclusive fitness is not important after all. If we are back at measuring the number of offspring, what is the point of it? Why should we be concerned with the effects of non-offspring relatives when offspring relatives seem to be the only ones that count? This is not what we should conclude as we can see from looking closely at what Hamilton did show.

Hamilton demonstrated how a trait could spread through its effects on the reproductive success of animals other than direct offspring. He emphasized that this was not a new sort of selection process but one which operated in essentially the same way as ordinary 'parental care' selection. In both cases the 'costs' and 'benefits' are paid in the same currency – loss and gain of adult offspring. Even with parental care, there is cost (to future offspring) of nurturing present ones.

But genotypes that never produce adult offspring cannot be successful. Hamilton showed that one of the ways that genotypes can increase their production of adult offspring is to have some members of that genotype diverting their efforts from reproducing themselves to helping the reproduction of other members of the same genotype. Wanting to express this idea not in terms of genotypes and gene frequencies but in terms that could be applied to individual animals, he devised 'inclusive fitness'.

Whether a genotype can increase its production of adult offspring in this way will depend on various contingencies. If one brother has a very good territory and is good at reproducing, and a second is underweight and has a poor territory, more adult offspring may be reared if the second weaker brother spends his time chasing predators away from nieces and nephews than if he attempts to rear a brood of his own. The genes for helping brothers under such circumstances will spread. On the other hand, if the brothers both have good territories and good chances of reproducing themselves, helping a brother at the expense of their own reproduction may be less advantageous than reproducing themselves.

Contained in those last two sentences is the answer to a question that may have occurred to you as you read this chapter: *If aiding relatives is so advantageous, why is sibling care less common than parental care?* Between full siblings, the coefficient of relatedness, r, = 0.5, just as it is between parents and offspring. On genetic grounds, then, there would seem to be the same advantage to caring for a sibling as caring for an offspring. Sibling care is, however, much less comon and to see why, we have to go back to Hamilton's original equation.

Remember that in Hamilton's equation, there were three terms, r,

B and C. B is the benefit to helping and C is the cost, both expressed in gain (or loss) in offspring. As the result of the last section, we are now in a position to see that these gained or lost offspring are best seen as offspring of a given genotype. The equation simply says that relative-helping will evolve when the gain in extra offspring as a result of helping outweighs the number lost as a result of helping. The same equation also tells us why parental care is so much commoner than sibling care.

We are considering the case where r (between siblings and between parents and offspring) is the same and so whether the result is parental care or sibling care must depend on the balance of B and C. Sibling care will occur when C (loss to own offspring as a result of aiding relatives) is low and B (benefit to other relatives) is high. In other words, if an individual has a poor chance of reproducing himself (perhaps there are no more territories or nest sites available so the cost of not even trying to reproduce is low), he may gain genetically by adding to the survival chances of his siblings, *provided* he can really benefit them. There is no point in laying down your life for your brother if your brother would be perfectly all right without you!

Only if a real benefit can be given to the sibling will sibling care have any chance of evolving. In bicoloured wrens, for example, a single helper can dramatically increase the chances of sibling survival if it stays and helps its parents. Parents raising chicks on their own produce an average of 0.4 offspring a year, whereas parents with a helper produce an average of 1.3 offspring a year (Austad & Rabenold, 1986). A single helper can therefore be said to really contribute to the survival chances of its siblings and on average will be responsible for 0.9 of an offspring, apparently through warding off ground predators.

So as we can see, the conditions for B and C favouring sibling care rather than parental care will not be all that common. There has to be some reason why trying to be a parent on one's own is going to be disadvantageous. Shortage of territories or a particularly harsh environment might be some of them. And there has to be some real benefit that can be given to the siblings over and above what they would experience if they were looked after by their parents alone. This will be far from common. Termites, where the king and queen carry out no parental duties and leave all the care of the next generation to their older offspring, provide one clear example. And in birds help can be given in the form of bringing food for the young and mounting guard duty at the nest while the parents are away. In mammals, on the other hand, the retention of the embryo inside the mother's body and her provisioning of it with milk, leave far fewer tasks that can be carried out by anyone else. Both paternal care and sibling care are much rarer in mammals than in birds as a consequence.

In addition to the relatively infrequent favourable balance of *B* and *C* that is needed for sibling care to evolve, there are other reasons why it is the exception rather than the rule. If the mating system is polygamous then sibling relatedness will be low. Only with monogamy will genetic relatedness to siblings be as high as that to offspring. There also has to be overlap of generations and parents must be still actively reproducing at the time when their first offspring could reproduce themselves. The required high degree of relatedness and the ensured supply of siblings to care for do not occur in all species and go a long way to explaining why sibling care is not all that common.

There are two important points to notice about degree of relatedness itself, though. Of course, a high r may help to tip the balance in favour of sibling care, but care of relatives could theoretically occur even if r were low. Suppose an animal were faced with an evolutionary choice between zero chances of reproducing itself and caring for a distant relative. Caring for a distant relative is genetically better than nothing so in the right circumstances, relative-helping could evolve even with a low r. It all depends on what the alternatives are.

Secondly, even a high degree of relatedness is not necessarily the way to genetic bliss. The social Hymenoptera (ants, bees and wasps) are often held up as an example of sibling care brought about through a high degree of relatedness (the haplodiploid mating system giving sisters a particularly high $r = 0.75$). Of course, the high degree of genetic relatedness the workers have to other sterile workers is not what matters since nothing is gained genetically by helping a sterile sister except in so far as that sterile sister may in future help a fertile sister. What matters is the similarly high degree of relatedness the workers have to the much smaller number of their fertile sisters, the young queens of the next generation, and it is the small number of those young queens that should make us pause for thought. There are far more workers in most social insect colonies than there are young queens. This means that each worker is not responsible for the survival of one young queen but has to share *B*, the benefit or genetic gain, from helping a young queen, with other workers. To say that sibling care evolves in ant colonies because of the genetic advantage to caring for a reproductive sister ($r = 0.75$) over an offspring ($r = 0.5$) is therefore an oversimplification. Sibling care evolves because the benefit of being partly responsible for caring for a sister ($r = 0.75$) outweighs the alternative of trying to reproduce

(r = 0.5) and being unsuccessful. One whole reproductive sister (a young queen) is not traded for one offspring, but part of a sister for an offspring with negligible chances of survival.

All of these complications, including inclement environments, shortage of territories and bits of sisters are in fact included in Hamilton's equation, which can cope with even more bizarre outcomes, like killing and eating relatives. Although aiding relatives is sometimes favoured for the reasons we have already discussed, the powerful mixture of r, B and C can also conspire to produce murder and cannibalism. If by killing a brother, an individual so increases his chances of survival that he lives to have many times more offspring than he would have done had he not killed his brother, then brother-killing will be favoured. If B (increase in own reproduction) is greater than C (loss of brother) then own reproduction will win, despite r being a respectable 0.5. This seems to occur in herons and egrets, where siblicide in the nest can be quite common, brothers and sisters can be stabbed and thrown out of the nest but the gains are considerably more food for the sibling that remains (Mock, 1984).

If you are disconcerted by the way in which B and C are sometimes expressed in terms of genetic advantage or loss to self and sometimes in terms of the same thing for relatives, the solution is to stop yourself from slipping back and thinking in terms of individuals (self or relatives) and keep thinking instead in terms of what we have seen inclusive fitness is really about, namely, gene frequency. The level at which natural acts is the gene and not the individual. What matters, in the end, is whether genes get copies of themselves through to the next generation. Some do it by making the body they are in better at surviving and reproducing. Others take a more indirect route and increase the survival chances of other bodies that contain copies of themselves, or at least have a fair chance of doing so. This chapter has basically been about what makes genes take one route or the other into the next generation. It is clear, however, that in order for the indirect route to work at all, genes have to have some means of effectively recognizing other bodies that contain copies of themselves.

Genes, bodies and recognition

In theory, there are two ways in which genes could direct the bodies they are in to aid copies of themselves in other bodies. One way is to use degree of relatedness, which is not foolproof because even close relatives may not possess the gene in question. *r* gives only a probabilistic estimate as to whether an individual body has a gene in common with another body. Thus a brother-helping gene that causes its bearer to sacrifice its own reproduction by aiding a brother might be unlucky. The helped brother might not have the brother-helping gene, so the gene may not, in this instance, increase the number of brother-helping genes in the next generation. It does so with a probability of only 0.5. On the other hand, there is much more chance that a brother (r = 0.5) has the brother-helping gene than a randomly chosen member of the population, so even though the probability that the brother has the gene is not 1 but 0.5, brother-helping genes may still propagate themselves, if they help enough brothers.

But think of how much better a gene could do if, instead of using the probabilistic label of relatedness, it had some means of picking out that sub-set of its relatives that definitely had a copy of itself. If it could pick out those brothers that had a copy of the brother-helping gene and ignore those that did not, it could spread much faster because it would not be wasting its body's effort on helping other bodies that did not contain a copy of itself. Hamilton produced the concept and R. Dawkins coined the term of 'green beard genes' to describe such genes. The idea behind green beard genes is that, in order for a gene to benefit a copy of itself in another body, it must have some way of recognizing the presence of that gene. One obvious way to do this is to use some unmistakable effect that the gene has, a unique flag or label that identifies its presence. Hence the somewhat absurd notion of a green beard. But in order to benefit a copy of itself in another body, the gene must not only recognize the flag, it must direct care to flag-bearers and it should also produce the identifying flag in its own body so that other green-beard carriers will in turn direct aid to it.

This requirement – that a single gene can identify, direct aid and signal its own identity – has led some people to dismiss green beard genes as very unlikely to occur in practice. (And unless it was a single gene, the whole system would be open to cheating: if different alleles were involved, a given individual might possess the gene that gave its bearer the green beard but not the one that delivered aid to other green-bearded individuals and so get the benefits without paying the costs.) However, it is not impossible to think of cases where it might occur. David Queller suggested that a gene that gave a nestling a particularly big mouth might in effect be a green-beard gene. Big-mouthed individuals would get more food from their parents than their small-mouthed siblings. The big-mouth gene would therefore have a more damaging effect on those of its siblings that did not possess the big-mouth gene than those that did. While this would be differential lack of harm rather than actual benefit, it would still be an example of a gene having a different effect on copies of itself than on its allele, which is the essence of green beard selection.

Whether or not green beard genes actually exist, they are a very important thought experiment because they help us to understand how inclusive fitness really works. If you can understand why green beard genes (that aided only individuals that definitely possessed copies of the green beard gene) would spread through a population faster than ordinary relative-helping genes (that aided relatives with only a probabilistic chance of having the relative-helping gene themselves), then you have almost certainly understood Hamilton's theory. You will also have made the switch from thinking on an individual level to thinking on a gene level.

From this, we can see that genes do not direct their bodies to aid relatives because those bodies 'share genes' in general. They direct their bodies to aid relatives because those relatives have a high chance of possessing the same relative-helping gene and that gene will under some circumstances be better represented in the next generation than a gene that did not aid relatives. The fact that the bodies of those relatives have other genes in common, such as the same eye colour or the same hair colour is irrelevant. What matters to the spread of the relative-helping gene is whether the relative-helping gene is present or not. If it could, it would help just those relatives (and even non-relatives) that had it and ignore the rest, however many other genes it had in common with its own body. All relative-helping genes would be green beard genes if they could, even though in practice they seem generally to settle for being relative-helping genes and simply live with the fact that some of their effort will be wasted on non-carriers of themselves.

This would even work between parents and offspring. If a gene for increased parental care could detect which of its offspring did and did not have copies of that gene for increased parental care, it should ignore the ones that did not and concentrate on rearing the ones that did. In this way, its representation in the next generation would be greater than by making the body it was in into an indulgent parent that fed and nurtured all its offspring. Even here, it is gene selection that really counts.

One final word of warning. Despite all this long discussion, it does not follow that every time we see an animal directing some sort of aid to a relative we are necessarily witnessing an example of kin selection. As Grafen (1990c) and Barnard (1991) have pointed out, helping relatives may sometimes be an incidental by-product of some other process such as species recognition or simply social interaction. If it benefits animals to group together, for example, it may be that they tend to group with their relatives because relatives happen to be closest. But if the same benefits of group living could be obtained by staying close to members of the same species whether they were relatives or not, we would not be entitled to use this as an example of kin selection. The genes would be getting through to the next generation through the benefits of group living but they would not be doing so by sacrificing the body they were in for the benefit of other bodies in which there were copies of themselves. They would simply make the body they were in do the best it can for itself.

Conclusions

Despite what we might like to think, it is the possession or non-possession of a gene that makes it advantageous to help relatives or offspring, not that they share much of the same genome. Inclusive fitness is, as was stated at the beginning of the chapter, just a device for letting us think that what matters is individuals when what we are really seeing as we watch animals go about their business is the working out of the different strategies by which genes get themselves into the next generation.

Further reading

The papers by Hamilton (1964) and Grafen (1982, 1984, 1991), read carefully, will set you on the right lines. R. Dawkins (1979) points out some of the many misunderstandings that have arisen about kin selection. Grafen (1990) and Barnard (1991) discuss problems of kin recognition.

4

GENES AND INNATE BEHAVIOUR

Grey Wagtails (*Motocilla Cinerea*) chicks
Photograph by Mike Amphlett.

The idea of genes getting themselves into the next generation by making the bodies they are in behave in certain ways may sound innocuous enough but has in fact aroused intense opposition. It has given rise to accusations of 'genetic determinism' (Gould, 1978; Lewontin *et al.*, 1984). The idea that behaviour has evolved and that there might be genes for behaviour in the way that there are genes for other characteristics is taken as synonymous with the view that genes control all we do.

It is impossible to dismiss this accusation by the simple, if correct, reply that this is not at all what is implied by the theory of natural selection. The criticisms have gone on for so long and have such a hold over many peoples' imaginations that a more explicit rebuttal is called for. Indeed, they have surfaced recently with renewed vigour in the face of the suggestion that there might be a 'gene for' homosexuality (Hamer *et al.*, 1993). In this chapter, we will tackle the task of explaining what it does and does not mean to say that there are 'genes for' behaviour. Then, having cleared up this definitional problem, we will look at the evidence for the existence of such genes. This will enable us to evaluate whether they possess the sinister, dominating properties that have sometimes been attributed to them. We will also be in a position to decide what is and what is not meant by calling behaviour 'innate'.

'Genes for' behaviour

Adaptation, we have seen, implies that natural selection has favoured one type of animal over another of a different colour, shape, or behaviour. If we wish to study adaptation, therefore, we have to search for these different types of animals and see why one does better than the other. Very often the original types are not available for us to study for the simple reason that natural selection has already occurred and eliminated all but a few. So studying adaptation often means adopting less direct approaches as we saw in Chapter 1, but all approaches involve making some kind of comparison. We compare the evolutionary success of one kind of animal with that of another.

Comparison dogs all studies of adaptation. The word implies comparison – of something, with something. Natural selection cannot operate unless animals show differences between which selection can occur. And we can go further: natural selection cannot work unless those differences are inherited, unless, that is, the traits which gave their parents the edge over their rivals are handed on to their offspring.

Natural selection, then, implies not just variation but genetic variation of some sort, however slight. This is not to say that this genetic variation exists now. Natural selection may have eliminated the variation that once existed. But we assume that some genetic variation must have existed in the past. There must once have existed black-headed gulls that did not remove eggshells and lost their chicks to predators as a consequence, even if such gulls are not to be found nowadays. We assume that these gulls must have been genetically different from the more successful eggshell-removing types. Otherwise we would be wrong to call eggshell removal an adaptation or to imply that natural selection had been at work.

So, a belief that behaviour has evolved in this way does indeed commit us to an assumption about genetic variation. We assume that there either are or there have been in the past individuals that differed from one another genetically with respect to behaviour. It is in this sense, and in this sense only, that we use the phrase 'gene for' behaviour. Just as 'adaptation' is correctly rendered as 'adaptive difference' (between two types), so 'gene for' can be replaced by 'genetic difference' (also between two types).

All that is implied by using a phrase like 'gene for helping a relative' is that there is some genetic difference between animals from a genotype that helps relatives and those from a genotype that does not. One genotype may have a slightly greater tendency to regurgitate food, for instance, when arriving at the nest. It does not mean that there is a single gene which, by itself, controls all the animals' interactions with relatives. It simply means that a small genetic difference could be responsible for making an animal into a good regurgitator or a bad regurgitator, perhaps through affecting the level of hormone secreted by a particular gland. Nor does it imply that all possessors of the 'gene for' helping relatives will necessarily help relatives under all circumstances. As we saw in the last chapter, for example, the 'gene for' sibling care in social insects is expressed in workers but not in young reproductives.

This idea of a 'gene for' a difference that may or may not be shown is not, incidentally, in a new meaning of a 'gene for' something. It is precisely what 'ordinary' geneticists, studying height in pea plants or eye colour in fruitflies also mean by it. To talk about a 'gene for' tallness in a pea plant does not mean that just one gene is responsible for the entire growth of the plant. Many genes will be responsible for that. But one or a few genes can be responsible for the *difference* in height between tall and short plants and in this restricted sense can be regarded as a gene or genes for tallness. And not all pea plants with the 'tallness' genes will be tall. Plants with 'tall' genes that are grown in poor soil will turn out to be short despite their genetic inheritance.

What is implied by the idea of 'genes for' behaviour is that some part of an animal's genetic material is responsible for bringing about alterations in the structure of that animal, which in turn affects its behaviour. This is surely not, when put in this way, such an outrageous claim. We know that genetic variation exists and has been documented for a large number of morphological characters. We are quite used to it for blood groups, coat colour, milk yield, and a host of other things. Is there also evidence for genetic variation in behaviour?

THE GENETICS OF BEHAVIOUR

There is, in fact, very good evidence now of this, of a sort very similar to that previously obtained for morphological characters. This is hardly surprising. Behaviour results from the interaction of sense organs, nervous system, muscles, and other parts of the animal's body. Variation in any of these might be expected to affect behaviour as well. Male crickets (*Teleogryllus*), for example, attract females by singing (produced by scraping one wing rapidly over another), and an alteration in the pattern of nerve impulses going to the wing muscles completely changes the nature of the song. Too many or too few nerve impulses or even just the right number of nerve impulses at the wrong times makes the wings produce a song that female crickets simply do not find so attractive (Bentley and Hoy, 1972).

Genes in the male are responsible for affecting the number and patterning of nerve impulses that go to the wings. Crossing two species that have different songs results in males that have songs intermediate between the two. The pattern of nerve impulses to the wings as they sing is also intermediate. So, genes affect nerve cells which affect wing muscles which affect the song produced which in turn affect the attractiveness of the male to females. The route from genes to behaviour is long but it is traceable, at least in this case.

In other species of cricket, the amount of singing a male does is also affected by his genes. In a species called *Gryllus,* Cade (1981) showed that there were two sorts of male. Caller males sit and sing for females. Satellite males sing much less but intercept the females on their way to the singers. Caller males have the benefit that the females know where they are but are much more likely to be parasitized because their parasites, like the females, use the song to locate them. Satellite males with the advantage of being less parasitized coexist side by side with singers, the balance of advantages and disadvantages between them being apparently equal.

Breeding experiments show that the difference between callers and satellites appears to be genetic. Breeding from caller males gives rise to sons that sing a great deal, whereas breeding from satellite males gives sons which tend to be silent for most of each night.

In many of the genetic studies of behaviour, however, the situation is not simple as appears from these examples. There is often an additional complication in that the environment can be shown to contribute substantially to the observed variation between individuals. If a 'satellite' male finds itself on its own, with no other calling crickets within earshot, he will start calling. In other words, although he is genetically a non-caller in an environment in which other males are calling, he can change his behaviour in a different environment as the occasion demands.

This simply emphasizes what it means to say that there are 'genes for' calling behaviour in crickets. Natural selection does not demand that all variation should be genetic or that underlying genetic differences should be apparent all the time. What is needed for it to operate is that at least some part of the variation between individuals should be genetic, and that variation should be shown at least some of the time. The more of the variation that is genetic, the faster natural selection will operate, but even a small genetic difference between individuals, apparent only under certain circumstances, will in the long run be enough. The course of evolution will be altered as long as successful individuals can pass on at least some of their characteristics of success.

Many behavioural characters have now been shown to exhibit this genetic variation. Even something as complex as 'recognizing relatives' may have a genetic base. Sweat bees (*Lasioglossum*) live in burrows in the soil with the entrance to the burrow patrolled by 'guard bees', which allow only certain bees into the nest. Whether or not the guard bees allow another bee in is, quite remarkably, directly correlated to the closeness of their relationship with it. If a full sister tries to get in, she will almost certainly be allowed to, but aunts, nieces, and first cousins are likely to be prevented in direct proportion to their degree of relatedness. Even sisters that the guard bee has never met are allowed in according to their genetic closeness with the guard bee. Half-sisters with a different father are less acceptable than full sisters with the same father. These in turn are less acceptable than full sisters of parents that are themselves related and which therefore have an exceptionally close genetic affinity (Greenberg, 1979).

It seems likely in this case that genetic differences between individuals result in their having slightly different odours – the more closely related the bees are, the more similar they will smell. The guard bees perform their relative-excluding feat by turning away bees that have the wrong odour. The recognition of relatives is accomplished by a subtle mixture of genetic variation (in odour production) and learning (of familiar odours).

From these and other examples, we can see that it is quite plausible to postulate genetic variation in behaviour for a number of different traits, including complex ones such as learning ability and interactions with relatives. There clearly are genetic differences between animals in the way they behave and in this sense, there are many 'genes for' behaviour.

Genes, behaviour, and genetic determinism

Does the existence of genetic differences in (that is, genes for) behaviour imply that those differences are in some way fixed and impossible to get rid of? In many people's minds, this is what the very existence of genetic variation implies. Genetic traits are seen as ones that no one can do anything about. To concede that genes affect behaviour becomes equivalent to saying that genes determine behaviour. They become the controllers, dictating what animals do at every stage of their lives. The fallaciousness of this view can be seen by considering two examples.

There is a serious single-gene inherited disease called phenylketonuria. Children inheriting a copy of this gene from both of their parents are severely mentally retarded and usually die before reaching their teens. The gene works by preventing the formation of an enzyme, which, in normal people, converts one amino acid (phenylalanine) into another (tyrosine).

Phenylanaline accumulates in the brains of children with phenylketonuria and causes the tragic effects on their mental development. The trait is due to a single recessive gene and we even know the biochemical effects that it has on the body. It is, however, quite possible to mitigate its effects by means of relatively simple changes in the environment. If the disease is detected early enough in life, affected children can be fed on a diet containing no phenylalanine. The result is that phenylalanine does not accumulate in the brain and the children are normal or nearly so (Hsia, 1970). A genetic difference between normal children and children with phenylketonuria can thus be largely eliminated by manipulation of their environment. The children still have the genes for phenylketonuria, of course, but the effects of the genes are overcome.

A second example that illustrates the same point concerns the ability of mice to run through a maze to find food. The particular 'maze' concerned consisted of a series of ladders, tightropes, and so on, which mice of different genetic strains had to learn to negotiate in order to get food. Mice of some strains were much quicker at learning this task than mice from other strains and at first this difference appeared to be largely environmental and have little to do with genetics (Henderson, 1970). This initial conclusion was arrived at by comparing mice all reared in the same standard laboratory mouse cages. However, when the experiment was repeated, with mice reared in 'enriched' environments (cages with toys and things for them to climb on), the learning ability of all the mice improved. An enriched environment in early life seemed to make them all better able to cope with the maze. But what was particularly interesting was that the genetic differences between mice from different strains became more apparent too. When the mice were reared in enriched environments, mice from one strain showed themselves to be very much quicker at learning than mice from the other strain and this could be shown to be a genetic difference. This difference had not been apparent when mice of both strains had been reared in ordinary environments. Neither strain was particularly good at learning. It needed the enriched early environments to sort them out.

These two examples again serve to emphasize that genetic differences may show themselves under some circumstances but not in others. Genetic differences between animals may not always be apparent. They may need a particular environment to show themselves at all. They are certainly not fixed and impossible to eradicate. Of course, some traits, like which blood group you are, do seem to be fixed by the genes you have, but behavioural traits seem to be much more open to modification. Genetic differences in behaviour may be radically altered, mitigated, or even reversed by altering the environment. Genes contribute to the observed differences between individuals, in their behaviour as in other things, but their contribution is not sacrosanct. It can be lessened or enhanced in just the same way as the contribution from the environment can.

That should be a relief to anyone who has misunderstood this fundamental point. It is possible for us to believe in adaptation with its implication of underlying genetic variability without having to become genetic determinists. We do not have to believe in the dominion or dictatorship of the genes nor to see natural selection making animals (ourselves included) into puppets, manipulated by their genetic masters within.

Genetic variability is indeed a logical necessity for the theory of natural selection, and there is overwhelming evidence for its existence even though the details may not be known as fully in a given case as we might like. But genetic variability is nothing sinister, even though it has often been construed in that way. Such constructions are based, as we have seen, on a misunderstanding of what the science of genetics is about and what a geneticist means when he talks about a 'gene for' something.

This is not the whole story, however. The arguments about the relative importance of genes and environment in behaviour cannot be so lightly dismissed. It would take more than a discussion of simple genetics to resolve finally what has been called the 'nature–nuture' argument. People have argued for many years about whether any behaviour can be described as 'innate' or 'instinctive'.

The trouble is, they have used these terms in many different ways and so added a rich measure of confusion to the whole controversy.

INNATE AS 'GENETIC'

Firstly, there is 'innate' meaning the same as 'genetic'. The dictionary definition of innate is 'inborn' so it is intuitively reasonable to interpret this as meaning 'in the genes'. We have just seen, however, how very careful we have to be with the relationship between genes and behaviour. To talk about 'genes for' behaviour implies that there are genetic differences between one variety of animal and another. It does not imply that the genes are the only cause of the behaviour or that the genetic differences are fixed and immutable.

If we substitute 'innate differences' for 'genetic differences', then the same arguments apply. It is quite possible for differences between individuals to be genetic or innate in origin, as the examples we discussed showed. But it is a fallacy to think that those differences cannot be changed by the environment. The differences may lie 'in the genes' but they are then no more fixed than if they were 'in the environment'.

Now, clearly, it makes sense to talk about differences (whether genetic or environmentally produced) only if we are talking about more than one individual. There cannot be a difference between two animals if there is only one animal and the problem is that ethologists have used the word 'innate' to describe the development of behaviour in one single animal. So they must then be using the word in a different way altogether.

INNATE AS 'THE OPPOSITE OF LEARNT'

The pioneering ethologist Konrad Lorenz (1932, 1937) was very influential in championing the developmental view of 'innate behaviour'. He saw an animal's behaviour as divided up into 'instinctive behaviour patterns', which he believed to be based on inherited nervous pathways, much like reflexes, only more complicated. These instinctive behaviour patterns (*Instinkthandlungen*) were 'innate' or 'inborn'. An animal did not have to learn how to do them either by imitation of another animal or by trial and error on its own. It certainly did not have any kind of 'insight' into what it was doing.

In Lorenz's view, there was a clear-cut distinction to be made between innate and learned behaviour, which, in practice, it was possible to make by rearing a young animal in isolation (so that it could not copy another member of its species) and denying it any opportunity for learning on its own. A stickleback reared in isolation and performing the zig-zag courtship dance just like any other male stickleback the first time it saw a female or even a model of a female, would be an example of 'innate' behaviour – the genes telling the animal what to do.

How could Lorenz be so naïve, wrote his critics (e.g. Lehrman, 1953) to believe that behaviour can be neatly packaged into 'innate' and 'learned' components quite separate from one another? What about other environmental effects besides 'learning' (such as amount of sunlight)? What about subtle learning effects an experimenter might not be aware of? And, above all, how could anything be entirely innate, entirely determined by genes? Every animal needs an environment to develop in. At the very least it needs food, oxygen, and a certain range of temperatures. If these are absent, it cannot develop and there is no behaviour. The idea of any behaviour being 'innate' – entirely within the genome and unaffected by any environmental factors – was dismissed as quite meaningless.

There is an important sense in which these criticisms are valid. Whereas it does make sense to say that differences *between* individuals are entirely genetic, it does not make sense to say that the development of behaviour *within one* individual is entirely genetic. Genes need environments in order to build bodies which can then behave. All behaviour is therefore due to both genetic and environmental factors. That is a truism. It is hard to believe that the arguments had no more substance than a denial of this obvious, self-evident fact. In fact they had, and the controversies flared largely because of misunderstandings, on both sides, as to what 'innate' actually means.

Accepting the obvious fact that all behaviour is due to both environmental and genetic factors, there are two possible ways of still keeping the idea that some behaviour is 'innate'. There is some behaviour that can be called 'developmentally fixed' – all animals of a species do it in almost any environment they have been brought up in. Short of boiling the animal or putting it in an environment in which it simply cannot survive, we find that almost no change in the environment has any effect. The animal grows up and carries out the behaviour.

A good example of 'developmentally fixed behaviour' is shown by the *Teleogryllus* crickets we mentioned earlier. The males sing in a way that is characteristic of their species and different from any other species. Even massive environmental changes – rearing crickets in isolation, subjecting them to the sounds of other species, and so on – have no effect. The male cricket persists in singing his own species' song. Of course, in the future it may be discovered that there is some environment that does alter the song. Perhaps eating a particular sort of yellow fruit will have the effect of making the male sing a different sort of song. But that has so far not been discovered. The singing appears to be very 'developmentally fixed' and certainly not learnt from other crickets (Bentley and Hoy, 1972).

We could, on this definition, refer to the singing as 'innate', but it would be necessary to point out that this did not mean that the environment was not important. Of course, developing male crickets have to eat and breathe and have hundreds of other interactions with their environments. But it looks as though, within the range of viable environments for that species, precisely which of those environments the crickets happen to grow up in has no effect on the singing.

In this respect, the behaviour is different from, say, human speech, which is highly dependent on the precise acoustic environment a person grows up in. So, although it might perhaps be useful to put a little flag 'innate' on the cricket singing to indicate its developmental fixity, we might also have to be ready with a lot of disclaimers for people who thought we were denying any role for the environment at all.

There is, however, another possible meaning for 'innate', which is rather more subtle and particularly important for any understanding of the controversies surrounding the word because it is the one that Lorenz himself put forward in a revised (or clarified) form of his ideas in 1965. In the light of the many criticisms of his earlier papers, Lorenz reiterated what he saw to be the central point about the development of species-typical behaviour: that it is adaptive, that the animal in some sense 'knows' what to do without having to learn. The male stickleback 'knows' that rival male sticklebacks are red underneath; in Lorenz's view the important thing is to understand how it comes by this information. It might learn it or it may carry it in its genes. Lorenz saw no other way in which the animal could come by the information. The fact that the animal needs to have many interactions with its environment (food, oxygen, etc.) in order to develop the behaviour Lorenz saw as quite irrelevant because no amount of food or oxygen could give it the *information* that rival male sticklebacks are red underneath.

To decide whether aggression in male sticklebacks should be called innate or not, Lorenz therefore thought it was quite unnecessary to investigate the possible role of all environmental factors that could be thought of. It was necessary only to discover how the information 'rival males are red' enters the stickleback.

Rearing the fish in isolation from other members of its species (making sure that it cannot see its own reflection in the sides of a glass tank) and seeing how it performs the first time it ever sees another male stickleback would go a long way to settling this question. Attention is thus fixed on the role of particular environmental factors (in this case, exposure to fish of the same species) rather than on the whole range of environmental factors of the other, 'developmentally fixed' method. The only thing that matters here is whether the development of the behaviour is affected by alterations in the one crucial factor that could give information about what rivals look like. The cricket singing would also be described as 'innate' in this sense, but on the basis of the one finding that males 'know' what to sing without having to learn through hearing other crickets.

This view of Lorenz's has immense appeal. He wanted to emphasize the fact that so much of what animals do is adaptive the first time they encounter a particular situation. The world of animals is not a gigantic Skinner Box in which they gradually learn, by trial and error, what to do and what not to do. They come into the world 'equipped by nature' to behave in ways that are likely to help them survive and reproduce. And this, Lorenz maintains, demands an explanation.

There is an intuitive sense in which 'innate behaviour' is a special category. If an animal 'knows' what its rival, mate, or food looks like without previous experience, it is showing a different sort of behaviour from having to learn by being rewarded or punished. Or is it? Are the two sorts really so different? Even 'learning' is not writing on a blank slate. Animals are much better at learning some things than others. Chaffinches have to learn their song from other chaffinches in order to produce the full species-typical sound. But they will only learn certain sorts of sound – chaffinch song or elements of it (Thorpe, 1961). A chaffinch cannot be taught to sound like a canary even though it may have to learn to sound like a chaffinch.

Such 'constraints on learning' chip away at the distinction we might intuitively want to make between 'innate' and learning behaviour. There are also examples of 'innate' behaviour being modified. Laughing gull chicks have an 'innate' tendency to peck at their parent's bill in order to make them regurgitate food. They do not need to learn what to peck at. They emerge from the egg with information about what a parent gull looks like even though the information is rather crude. They will beg particularly vigorously to long thin red knitting needles because these have many of the characteristics of the parent's bill (Hailman, 1967). The initial innate response, however, is subsequently modified by experience. The chick learns, as it grows, the finer aspects of what its parent looks like and becomes much more discriminating in what it shows the begging response to. Older chicks will beg only from real gulls or models that look very like the parent laughing gull.

The distinction between 'innate' and 'learned' is further eroded by the song sparrow (Slater, 1989). Song sparrows reared in isolation develop perfectly normal song-sparrow song. But some sparrows deafened so that they cannot hear themselves singing, develop only a very rudimentary type of song. The birds need to hear themselves singing. They practise until the song sounds right, 'right' meaning that it sounds like a song sparrow. They have an innate idea of what the song should sound like but they need the environmental feedback of hearing themselves sing – an 'innate schoolmarm' as Lorenz called it.

So, the intuitively reasonable distinction between 'innate' and 'learned' based on 'sources of information' is difficult to maintain, at least if we insist that it must be hard and fast. Despite this, the word 'innate' or at least the concept refuses to disappear altogether. It still appears in the literature, sometimes modestly clothed with quotation marks ('innate'), sometimes awkwardly as 'what we used to call innate'. People grope towards a category of behaviour, even though it may be a difficult one to define precisely because they want to emphasize one of the most remarkable features about animal behaviour – the fact that an animal may behave in a manner appropriate to its survival and reproduction the first time it finds itself in a particular situation.

A hard-line attitude would be to say that although that is an important point to make about animal behaviour, making it is hindered rather than helped by the label 'innate'.

But whether or not we use the label, there clearly is some behaviour that does have special properties. This behaviour develops in much the same way whatever the animal's early environment is like. No special experience is necessary for the animal to show appropriate, adaptive behaviour. The singing of *Teleogryllus* crickets is an example. A great deal of confusion has arisen, however, from the common belief that behaviour showing these properties of being unlearnt, adaptive, and 'developmentally fixed' must also, logically, be 'genetic'. Our next task is to show why this is not so.

'Developmentally fixed' does not mean 'genetic'

We have already seen that the development of behaviour in the individual should be distinguished from differences between individuals making up a group or species. Development can be followed in single animals. We can watch them grow up. We can try to isolate the factors that influence their development. But genetic differences between animals ('genes for' different characters) cannot be studied in single animals. There must be more than one for the idea of 'differences' to make sense at all.

This has an interesting consequence. It means that heavily selected traits – those for which natural selection has favoured only one genotype – tend not to exhibit genetic differences because only one successful variant may now remain. All individuals may be genetically similar for those traits because any individuals that were genetically different will have been selected against. Natural selection could have deleted them, leaving a genetically homogeneous population with, by definition, no 'genes for' those characters. As we have seen, this is not the same as saying that those characters are not influenced by genes. It simply says that there are now no differences between individuals that can be ascribed to differences in their genotypes.

Many of the traits described by ethologists as 'innate' fall into this category. Most or all members of the species behave in a certain way. They all remove eggshells or sing a particular sort of song, despite considerable differences in what they experience when they are young. The behaviour is 'developmentally fixed'. But, if there is little or no genetic variation, it would not be possible to say that there are 'genes for' the behaviour.

In this case of singing in male *Teleogryllus,* we have seen that this could be said to be 'innate' in the sense of being environmentally fixed in the face of many environmental manipulations. All crickets of the species sing in the same way. Measured within one species, therefore, there is no genetic variation in singing. Just looking at one species, we would find no 'genes for' singing behaviour.

If instead of considering just one species, however, we look at differences in singing behaviour *between* species, then we can identify 'genes for' singing. There are genetic differences in singing behaviour between *Teleogryllus oceanicus* and *T. commodus* as we saw earlier. Within one species, in other words, we can show that there is developmental fixity but the lack of genetic variation makes it impossible to identify any 'genes for' the behaviour. Only when we look at more than one species is it possible to find large genetic differences and conclude that there are 'genes for' the behaviour.

The apparent paradox between developmental fixity and genetics should now be beginning to resolve itself. Genetic differences (between individuals) are quite separate from developmental fixity (within an individual). To use the single word 'innate' to cover both or to assume that 'innate' can be equated with genetic, is a sure recipe for confusion, one that Lorenz's early papers unfortunately did a lot to create.

To find out whether the behaviour is developmentally fixed, we have to study the effects of different environmental factors on development of individuals. To find out whether there are genetic differences, we have to do breeding experiments or resort to other ways of finding out whether similarities of behaviour result from similarities of genotype. Even then, as we have seen, the answer will depend upon the range of genotypes we have chosen to study. The secret is to demand always two sets of facts – one about genetics and the other about development. If we do this, we will not go too far wrong. But if we are content with just one set, such as evidence that the trait is heritable under some circumstances and believe that this has told us anything about modifiability during development, we will be led, very surely, astray. It is the failure to recognize this distinction that is at the root of the confusion over the role of genes and behaviour. Showing that there is a 'gene for' a trait, be it homosexuality, learning ability, or the height of a pea plant means simply that there are genetic differences that appear under some circumstances. It says nothing at all about how the trait develops in the individual, whether it can be modified or whether there are some circumstances under which it will not appear at all. To find that out, we have to do a quite separate set of investigations.

Conclusions: should we talk about 'innate' behaviour at all?

It is small wonder that the word 'innate' has been the object of so much argument and debate. Two people arguing about it might each be meaning quite different things. One, thinking that innate meant 'entirely due to genes', might criticize his opponent for subscribing to a vacuous sort of genetic determinism. The other, meaning only that there are some things all animals of a species do without any opportunity to learn, might be rather taken aback at the vehemence of the attack on him. It would take a very long time to explain to each of them where their differences originated.

One possible solution would be to suggest that neither of them should use the word 'innate' at all and insist that they should explain, without once using that term, exactly what they each really meant to say. (We might even want to impose a similar 'explain-what-you-mean' condition on the phrase 'gene for'.) In other words, because the word 'innate' has been responsible for creating so much confusion, is there not a good case for abandoning it altogether? Although this would have some advantages, it would be a very retrograde step to abolish the word and then fall back on the bland statement that all behaviour is a complicated interaction of genetic and environmental factors (a complex nexus as it has been called). This would get us nowhere in understanding how behaviour develops.

It also obscures a very important fact. Animals do many things which they can have had no opportunity to learn how to do. They are born or hatch out of an egg and immediately behave in ways that help them to survive. Their behaviour, which makes them seek shelter or find food, is as much part of their equipment for survival and as much a product of natural selection as any scale or spot or feather. This we should not forget, even though 'innate', with its motley retinue of meanings, may be the wrong word to be trusted with such an important message.

FURTHER READING

Lewontin *et al.*'s book *Not in our genes* (1984) shows the kinds of criticisms that the idea of genes for behaviour has aroused. R. Dawkins (1982, Chapter 2) shows where some of the fallacies lie. Hailman (1967) and Gould & Marler (1987) show how inherited and environmental factors interact in development.

5

COMMUNICATION

Red Junglefowl cock (*Gallus gallus*). Photograph by Marian Stamp Dawkins.

Communication forms the fabric of animal social life. It is the way animals influence one another to come together in schools, flocks and herds as well as to space out and to defend territories. It includes the way that the sexes interact in courtship, rivals settle disputes without fighting and parents care for their young. In fact, looking at the way in which animals spend their time, it is striking how much of it they spend either influencing, or being influenced by, the behaviour of other animals – in other words, in some form of communication. It is not surprising, therefore, to find that people who study animal behaviour have given so much attention to the study of animal communication. What is surprising is that, despite this intensive study, the whole subject is extremely confused, largely because of the definitions of the various terms that have been used. While this was already true when the first edition of this book was written, the confusions have now reached monumental proportions, with leading theorists even disagreeing as to what should properly be called 'a signal' or 'communication'.

Why 'communication' and 'signal' should be difficult to define

At first sight, it is not at all obvious why there should be any difficulties with either of these terms. Both are words which are used frequently in ordinary language without any great misunderstandings appearing to result. But let us see what happens when they are applied to some of the most impressive examples of animals influencing each other's behaviour: flocking in birds and schooling in fish. In both cases, there can be such good coordination between the individual animals that the group seems to take on a life of its own. The flock or school can wheel and twist about, all the members sticking closely together and all seeming to turn at just the same moment. Here, surely, is communication, and of such a complex sort that we humans, looking in from the outside, may need slowed down film or videotape to see what is going on (Davis, 1980; Potts, 1984).

And yet, if we turn to almost any textbook on animal behaviour, we are given a very different picture. Animal signals, we read, have evolved so that they are conspicuous and exaggerated. They have become changed in the course of evolution so that they are more effective at altering the behaviour of other animals, a process that has been called 'ritualization' (Tinbergen, 1952; Huxley, 1966). Some signals are clearly like this. In the courtship of many bird species such as ducks, bright, conspicuous plumage in the male is combined with repeated stereotyped behaviour patterns to present the female with an eye-catching and unmistakable show.

But to insist that this is always the case when one animal influences another seems to fly in the face of the facts. Many animals clearly do influence each other not by obvious signals but by subtle tiny movements, particularly when they are moving as a group, so why do textbooks insist that conspicuousness and exaggeration are the chief characteristics of animal communication?

Most textbooks give a definition that appears at first sight to be quite straightforward to apply to animals. The definition is usually based on the idea that communication occurs when one animal's behaviour can be shown to have an effect on the behaviour of another. 'Signals' are the means by which these effects are achieved. However, various refinements are then introduced to distinguish communication from other sorts of effects that animals may have on one another, and it is these refinements that cause the problems. For example, one common qualification that is made is that the effect one animal has on the other should not be called 'communication' if it involves direct physical contact but only if there is some 'economy of effort'. Cullen (1972) wrote, '. . . to a man the command "Go jump in the lake" is a signal, the push which precipitates him is not'.

In other words, achieving ends by force is not to be called communication, whereas influencing another animal by making a sound or a gesture should be. Dawkins & Krebs (1978) express this by saying that whereas force uses the individual's own muscle power, communication makes only minimal use of this, relying for its effect on the muscle power of other individuals. 'A male cricket does not physically roll a female along the ground and into his burrow. He sits and sings, and the female comes to him under her own power. From his point of view, this *communication* is energetically more efficient than trying to take her by force.'

But how much more efficient does something have to be than force before we call it communication? In some cases, the economy of effort is obvious. A bird gives an alarm call and the whole flock takes flight. Here a relatively small signal gives rise to the energetically far greater response of all the birds taking off and moving a considerable distance.

In other cases, the economy of effort, if it is achieved at all, is much less obvious. Red deer stags challenge each other for the possession of groups of females during the rutting season by roaring at each other and often one stag will retreat after a roaring match with no fight taking place at all (Clutton-Brock & Albon, 1979). Clearly there is some economy of effort here because roaring is less dangerous and less exhausting than overt fighting, but roaring is not effortless either. Stags that have been roaring at a high rate are temporarily exhausted, even if they have won, and they are unlikely to be able to fight seriously for a while afterwards. Even though the victor may not have used his muscles to push his rival away, he had to make his muscles work very hard (at roaring) before the rival eventually moves off under his own steam.

'Economy of effort' is introduced into the definition of communication to distinguish it from overt physical manipulation but it is clear that it does not enable any hard and fast lines to be drawn. Rather, there is a continuum with small inconspicuous, highly economical signals at one end and overt physical violence at the other. In the middle, and often seeming to be not far from actual physical violence, are the large, loud or conspicuous signals, such as the roaring of red deer, which are so exhausting to produce that the 'economy of effort' is almost lost.

A second common qualification to the definition of communication causes even more problems. It is introduced to deal with the fact that not all 'influences' of one animal on another appear to be real 'communication'. For example, if an insect that relies on camouflage for its protection suddenly moves a leg and, by this behaviour, influences the behaviour of its predator so that the predator comes and eats it, most people would not want to say that the insect is 'communicating' with its predator.

To distinguish this kind of 'influence' from true communication, the definition is modified once again so that only cases in which the influence from one animal to another is achieved by *specially evolved* signals or displays are to be counted as communication (e.g. Krebs & Davies, 1993). 'Specially evolved' signals are recognized by being exaggerated, conspicuous and very obvious (Tinbergen, 1952). Cases where an animal behaves and just happens to influence the behaviour of another are not communication. On this modified definition, a male duck showing a complex, stereotyped and gaudy courtship display is definitely communicating, whereas the unfortunate insect is not. The duck's display has been specially evolved to gain the attention of the female, whereas there has not been any evolutionary change in the insect case to make the leg movement more conspicuous or more likely to attract the attention of the predator.

Anxious to arrive at a definition of a signal that allowed for the inclusion of duck courtship but excluded insects being eaten as a result of their behaviour, most ethologists have therefore insisted that a 'signal' is not just any behaviour by one animal that alters the behaviour of another. Such behaviour is called a signal only if it has been specially evolved in some way – perhaps through being eye-catching or loud – to enhance its signal function. The 'specially evolved signal' refinement of the definition of communication seems, therefore, to be making entirely plausible and reasonable distinctions between what should and what should not be called 'communication'.

It also, however, brings itself into direct conflict with the equally reasonable first refinement of the definition of communication – that of 'economy of effort'. It is the attempt to apply both of these parts of the definition simultaneously that brings about the first of our definitional problems with communication. On the one hand, communication is said to occur when animals influence each other not by physical force but by 'economy of effort'. On the other hand, communication is said to occur only when animals influence each other by means of specially evolved, ritualized signals, which are recognized by being large and exaggerated and not at all economical. To put this conflict between the two parts of the definition more succinctly, the most 'specially evolved' signals are those that are least economical in terms of the effort needed to produce them, but it is 'economy of effort' which is supposed to characterize signals.

All this means that, very much against any intuitive idea of what the term may mean, wheeling flocks of starlings or shorebirds are not 'communicating' at all because they do not appear to be using conspicuous exaggerated signals.

There must be something wrong. There must be a way of defining 'communication' in such a way that it excludes the unlucky insect and includes flocks of starlings. If there is, it is not commonly used. Most textbooks stick to the 'specially evolved' criterion, thereby confining the whole discussion of animal communication of a subclass of large and exaggerated signals. Very little attention is paid to the more subtle and unritualized ways in which animals influence each other's behaviour. They have generally been left out, by definition, from the study of animal communication.

Now we might think we could get out of this definitional impasse by avoiding the word 'signal' altogether and discussing information transfer instead. If animals are influencing each other, even though by very subtle means, it would surely be reasonable to say that they were 'transferring information' even though it might not be appropriate to say that they were using conspicuous 'signals'. Surely dunlin and starlings transfer information even though their flocks may be coordinated by small unritualized movements? It seems, however, that if we take this apparently reasonable step, we find ourselves up against yet another definitional problem: the innocuous, neutral-sounding term 'information transfer' has unfortunately picked up two quite separate definitions. Its use thereby creates an even greater amount of confusion than ever.

The two meanings of 'information transfer'

'Information transfer' has both a technical and an everyday meaning. The technical meaning is taken from that used by communications engineers and their need to have a precise measurement of the amount of information being transferred down a wire, say, or beamed from a satellite. It seemed a very good idea to apply the precise formulae of engineering to animal communication as it held out the promise of being objective about a subject that is in constant danger of being misinterpreted by subjective views of what animals are doing. It certainly achieves objectivity, but not quite in the way people think it does.

'Information transfer' in the technical sense refers to an increase in the ability of an observer to predict what is going to happen next as a result of a given event. For example, suppose I knew that you were going to one of four cities – York, London, Glasgow or Edinburgh – but not which one. Then you tell me that you were going to a Scottish city. I would still not know exactly where you were going but what you had just told me would have made me somewhat better able to predict. My uncertainty about your destination would have been reduced from 1 in 4 cities to 1 in 2 (one of the Scottish ones). In this sense, you would have given me *information*. This information is measured for the sake of convenience (engineers' convenience, that is) not in decimal-based numbers, but in binary based ones (0 or 1) or bits. The number of bits of information can be worked out by seeing how many yes–no (i.e. binary) questions would be needed to get the right answer.

So if, as far as I knew, you were initially equally likely to go to any one of the four cities, the number of binary or yes–no questions I would have to ask you before you said anything would be two (I could ask whether you were going to Scotland or England and then, having established that, ask you which of the two appropriate cities there you were going to). After you had told me you were going to Scotland, I could simply ask whether you were going to Glasgow and then, whatever you replied, I would know where you were going. My uncertainty about your destination is reduced from two bits (two yes–no questions) before you say anything, to one bit when you tell me you are going to Scotland and to 0 when you say 'Yes' to 'Is it Glasgow?'

This guessing game has been made deliberately simple to give the idea of what it means to be able to measure 'information transfer' (Shannon & Weaver, 1949). It becomes slightly more complicated when the possible outcomes are not equally likely, if, say, I know from past experience that you are more likely to go London than anywhere else. But it does not matter to this discussion whether you are familiar with these complications or not.

What does matter is the idea that, in all cases, we measure the amount of information transferred by specifying what the possible outcomes are before and after the transfer has taken place. When this is applied to animal behaviour, it is done in a way that is slightly surprising to many people. What is measured is not how much information is transferred to *another animal*, as might be thought, but how much is transferred to a *human observer* looking at the two animals. Suppose animal A gives a signal; animal B then runs away. This happens a sufficient number of times that we are sure that this is genuine case of A influencing B. B seldom runs away unless A has just given the signal and every occurrence of A signalling leads to B running off. We might think we could measure the amount of information A has transferred to B, but that, of course, would be impossible. We have no idea what B thought before or after the signal – no means of guessing how much its uncertainty was reduced because we have no idea of what its uncertainty was, or is.

The only uncertainty we know anything about at all is that of the human observer who has been recording the behaviour of these two animals over a long period of time. He or she knows that B has a repertoire of some 32 behaviour patterns. He also knows that when A has just made the signal, one of these 32, running away, becomes suddenly much more frequent than before. If A has just signalled, the uncertainty about B's behaviour drops suddenly and it immediately becomes very easy to predict what it will do next. We may not know what information has been transferred from A to B, but we can measure that transferred to the observer. Before the signal he knew only that B would perform one of 32 possible behaviour patterns. After the signal, he knows it will be only one particular one. On the simplifying assumption that all 32 were, before the signal, equally likely, his uncertainty is reduced from five bits ($32=2^5$) to 0.

The technical meaning of 'information transfer', then, is synonymous with 'influence' of one animal over another. It covers what is going on between starlings in a flock, fish in schools, red deer roaring and the insect being eaten after moving its legs. If we note that insects are more likely to get eaten when they have just moved, then seeing an insect move when there is a predator about gives us information – we know that its chance of being eaten has gone up. We are in a better position to predict what is going to happen next and in that sense, the insect has transferred information to us.

The technical sense of 'information transfer' is therefore considerably broader than 'communication'. 'Information transfer' covers all cases where a human being can better predict the behaviour of one animal knowing what another has just done. It runs the whole gamut of influences that one animal may have on another, with no restrictions as to how this influence is achieved (specially evolved signal or inadvertent movement).

We may at this point be tempted to say that this is a rather odd way to use 'information transfer', one that runs counter to ordinary, everyday use and also to the ethologists' definition of communication as occurring only if there are special signals involved. This is certainly true and that is precisely why the term 'information transfer', introduced with the intention of clarifying and simplifying the study of animal communication, has in fact had the opposite effect. It seems so restrictive to say that we can only talk about our own uncertainty being reduced when animals do seem to transfer information to other animals and not just to the human beings who happen to be looking on. Wanting to make this point, many ethologists use 'information transfer' not in its technical sense, but in a sense much closer to the ordinary usage of the term.

'Information transfer' in this other sense does refer to animals giving information to each other. The red deer stags we discussed earlier appear to transfer this sort of information about their fighting ability to each other (Clutton-Brock & Albon, 1979). As we saw, the stags assess each other for strength and potential fighting ability. But if we were to say, in a colloquial sense, that the stags were 'transferring information' about their fighting ability, we would be using the term in a quite different way from the technical sense we have just discussed. We would be implying (in a colloquial sense) that a stag was transmitting information that if he *were* to fight he would be able to fight hard and well. We would not imply that he was signalling that he was *going to* attack, which would be the technical, 'ability to predict the future' meaning of information transfer.

Prediction of future events, which is the key to information transfer in the technical sense, is not what is meant by this other more colloquial meaning – a difference which has led to no end of misunderstanding. And as if this were not confusion enough, Richard Dawkins & John Krebs (1978) then suggested that we should drop the term 'information transfer' from our descriptions of animal communication altogether and talk instead about 'manipulation'.

Manipulation and the dangers of being manipulated

Dawkins and Krebs argued that although in some circumstances it might be appropriate to describe animal signals as transferring information, in many other, perhaps most, cases there would be such a conflict of interest between signaller and receiver that it is more accurate to describe the signaller as attempting to 'manipulate' the receiver rather than just inform it. For example, if two animals are in conflict over a piece of food or a mate, each combatant would gain if it could 'manipulate' the other into retreating. A poor fighter that gives a threat display and causes its stronger rival to flee without a fight would have successfully manipulated it. An angler fish that dangles a worm-like bit of skin in front of a small fish and catches it because the smaller fish snaps at the 'worm' can certainly be said to have carried out a successful manipulation of its prey. In both cases, if information has been transferred, it is most definitely false.

Such manipulation of receivers by signallers (and well-documented cases of mimicry show that this can and does happen) naturally sets up counter-selection on receivers *not* to be manipulated. They should respond only to signals that, on average, convey 'true' information. In fact, we should expect to find successful manipulation occurring only very rarely. Batesian mimics, for example, are rare relative to their models so that predators still benefit on average by avoiding brightly coloured prey even though they may be occasionally 'manipulated'. Similarly, angler fish are relatively rare in the lives of small fish which, consequently, usually benefit from snapping at worm-like objects.

Zahavi (1975, 1977) proposed a theory of how receivers might discriminate 'true' from 'false' information and so avoid being manipulated. He suggested that signals which were costly to produce – those that took up time or energy, for example – were the most likely to carry 'true' information because they would be the least easy for a signaller to fake. Roaring in red deer is a good illustration of this. Roaring is an exhausting activity that can only be carried out at a high rate by stags that are genuinely fit and healthy. Those whose body condition is poor are simply unable to produce a high number of roars per minute. By using the number of roars per minute as the signal of fighting ability, then, a receiver could gain information that is more likely to be 'true' than if it responded to a signal such as tongue flicking, foot stamping or a low intensity sound that took little effort, all of which could also be given by a stag in poor condition.

Zahavi argued that in any situation where there is a conflict of interest between signaller and receiver – between rivals, between mates, between predators and prey, for example – there will be selection on receivers to respond only to signals that are large or loud or in some way costly for the signaller to produce, as this will be the best way for receivers to avoid the ever-present danger of being manipulated. There is no 'economy of effort' here. On the contrary, as Zahavi expressed it, the signal that is responded to by the receiver becomes a 'handicap' for the signaller, the truth or honesty of the information it carries being guaranteed by the cost of giving it.

All this sounds suspiciously like 'ritualization' – the name given by the early ethologists to the process by which signals became more adapted to their signal function during the course of evolution, usually by becoming more conspicuous or distinctive. But it is in fact a quite different explanation of why signals should change over the course of evolution. Whereas 'ritualization' meant various forms of exaggeration so that signals become better adapted for information transfer, Zahavi's idea is that signals become large as an adaptation for reliability or truthfulness. Which is right?

Cost, honesty and handicaps

Zahavi's idea that the honesty or truth of a signal comes from the cost paid by the signaller in giving it has had – to put it mildly – a checkered career. When it first appeared in 1975, it was strongly criticized by Maynard Smith (1976) who argued that in its original form it simply would not work. Zahavi had originally argued that a male animal might indicate his true health and strength to a female by growing a large ornament, such as antlers or a display of feathers, that was so large and cumbersome that it increased his chances of, say, being caught by a predator. If he survived despite the 'handicap' of the ornament, he would therefore be signalling that he must be exceptionally strong and healthy.

Maynard Smith replied that he would in fact be no stronger and no healthier than any other male since any male that had survived to that point would be as good as any other. If two runners compete in a race and both come in at exactly the same time but one is carrying a sack of coal on his back, we are tempted to say that the one with the 'handicap' of the burden must be the better runner in order to have done so well. And so he probably is, but only if, for the next race, he can throw off the sack of coal and compete unhandicapped, when we would expect him to win. However, suppose that instead of a sack of coal, this runner has an inherited physical handicap that gives him a permanently curved spine so that he is practically bent double. If he comes in at exactly the same time as the able-bodied man, we may be impressed by his performance but we cannot say that he is a better runner. He is as good, but no better. A female that chooses a strong 'handicapped' male with a long tail or large antlers is similarly choosing a male that is as good as but no better than a weaker unhandicapped male with a shorter tail or smaller antlers since both have survived equally well to date. If the handicapped male then passes on his handicap to his offspring, his sons may benefit from inheriting his strong constitution but they will also suffer from inheriting his debilitating burden. As far as the female is concerned, her sons will not have gained anything by her choice of a handicapped male as their father (although her daughters, inheriting his constitution but not his handicap, might benefit).

It subsequently became apparent, however, that although this original version of the handicap theory would not work, a modified version just might (Zahavi, 1977; Andersson, 1982). If the development of an ornament such as a long tail were dependent on the health and strength of a male during his own lifetime, then the tail might be used to signal true information about these qualities in the male. This idea still makes use of Zahavi's fundamental idea that signals cost signallers something, but makes the cost that is paid dependent on phenotypic state. If growing a long tail costs the male a considerable amount (by diverting resources that might otherwise go to muscle building, for instance), then only males that were so healthy and strong that they could 'afford' to pay this cost would grow long tails. The rest would grow shorter tails and females could then discriminate male quality on the basis of tail length and be sure that they were getting 'honest' information.

There is a seeming paradox here: males signal how strong they are by growing a tail that makes them less strong, but this is what Zahavi meant. He used the analogy of a wealthy man signalling how wealthy he was. He could do this by giving a huge banquet that cost him a lot of money. In other words, he would show how much money he had by using up some of it and thus making himself poorer. This spending of wealth to indicate wealth would, therefore, have a built-in check against cheats because cheats who were pretending to be rich but were in fact poor would bankrupt themselves if they tried to do what a genuinely rich man could do without making anything other than a minor dent in his capital.

The vindication of Zahavi's near-paradoxical idea came in 1990 with the publication of a model by Grafen (1990a, b) who translated wealth into tail length and showed that, at evolutionary equilibrium, tail length would be an honest indication of male quality. His model, like Zahavi's modified theory, assumes that a long tail costs any male growing it in terms of something of importance to him, such as health, vigour or ability to combat disease. It also assumes that a tail of a given length would be less costly to a genuinely healthy, good 'quality' male than to a poor quality male. Any male that grew a tail that was longer than his constitution could afford would therefore, like a poor man trying to impress his girlfriend by taking her to a more expensive restaurant than his bank balance could stand, lose out in the end. At equilibrium, all males grow tails that genuinely reflect their underlying quality. Tail length is therefore a signal of 'true' information about health and strength.

This might seem as though all honest signals were handicaps. However, as we will now see, this is not necessarily true, and to understand why we have to go into yet another and in many ways even more confusing controversy about animal communication.

ARE ALL SIGNALS HANDICAPS?

Maynard Smith (1991) made a major contribution to the study of animal communication by distinguishing two kinds of signal or, as he put it, two kinds of workable handicap. On the one hand, he said, there are the 'conditional' handicaps which we were discussing in the last section, conditional, that is, upon the state of the animal giving them. Grafen (1990a, b) called them 'strategic choice' handicaps because each animal has an evolutionary choice about which size handicap to develop but the stable strategy is to choose the size that accurately reflects quality and health.

The other kind of handicap Maynard Smith called a 'revealing' handicap. A revealing handicap is not particularly costly to produce but it simply cannot be given unless the signaller genuinely possesses the quality it is signalling about. For example, female toads are attracted to males with low frequency calls (Davies & Halliday, 1978). They benefit from mating with large males because large males have more sperm. Large males have lower pitched voices and the pitch of their croaks is a true indication of body size because it depends on the length of the vocal cords, which in turn depends on the size of the animal. Small toads therefore cannot make low frequency sounds, not because it would be too costly for them if they did (as would happen with a strategic choice handicap) but because it is physically impossible for a low frequency sound to be made by small vocal cords. The pitch of the call simply 'reveals' the true body size of the caller without costing it anything extra. Large body size may, of course, have its costs such as delayed reproduction but the costs are not specific to signalling in the way that strategic choice handicap costs are.

The distinction between a conditional (strategic choice) handicap and a revealing handicap is therefore partly in how costs are paid and partly in how the costs are related to the quality being signalled about. With the conditional handicap, the signal itself is costly and, more than this, it is costly in such a way that the signaller reduces or *uses up* the quality being signalled about. The profligate rich man we were discussing earlier demonstrated his wealth by spending money, not by, say, running round the park. To demonstrate wealth, he spent wealth whereas to demonstrate athletic prowess he would have to do something that 'handicapped' or 'used up' his physical strength. Similarly, if a male gives out information about his health and freedom from disease through a signal, and that signal is to be an honest, conditional handicap then that signal should make the male less healthy. If an animal is signalling about its fighting ability, then it should use a signal that makes it less good at fighting. The conditional handicap that evolves in any given situation, in other words, is related to the quality being signalled about – not any old tail, but a tail that damages ability to resist disease, if that is the basis of female choice. Not any old hump but a hump that interferes with ability to fight if that is the basis of a rival's assessment.

With a revealing handicap, on the other hand, the signal itself may be costly (in that it might, say attract the attention of predators) but it does not *use up* the quality being signalled about. So if a male were to use a signal that 'revealed' his fighting ability or health, he would not necessarily become worse at fighting or less healthy. Nevertheless, the revealing handicap might still convey honest information about these qualities, if the signal were necessarily correlated with, say, body size and body size was in turn a good indicator of fighting ability. Being a large size does not 'handicap' a male's fighting ability, but it may reveal it.

We can see the distinction between conditional and revealing handicaps more clearly if we look at two specific examples. Take one example we have already referred to several times – the roaring of red deer during the rutting season. If we interpret this as a conditional handicap, we would have to say that the deer handicap themselves by roaring, that is, they indicate how good they are at fighting by reducing their ability to fight, albeit temporarily, while they get their breath back. If we were to interpret this as a revealing handicap, on the other hand, we would say that only genuinely good fighters could roar at a high rate and that roaring was not costly in terms of ability to fight – it simply revealed fighting ability in an unambiguous and uncheatable way.

Or take the peacock's so-called tail (in reality, a fan of back feathers). If it is a conditional handicap, signalling health and freedom from disease, then growing a tail would be damaging to the male's health or immune system: males should make themselves less healthy by growing them. But if the tail is a revealing handicap, then the growing of the tail should not damage the male's health. Rather, only healthy males would be able to grow impressive tails, which 'reveal' their healthy, parasite-free state because long iridescent feathers are an inevitable by-product of being in good condition.

At this point, the perceptive reader may well have become worried, as well he or she might. Are there in fact any examples of conditional handicaps? Do animals really 'use up' qualities, make themselves less healthy or less good at fighting just to show how even healthier or good at fighting they would have been had they not been signalling? Grafen's models certainly show that this seemingly paradoxical idea can work in theory, but does it apply to real animals? Even Grafen admits that his model may have limited applicability to the signalling systems with which biologists are familiar. If his model and the whole notion of a strategic choice or conditional handicap were nothing more than an interesting thought experiment, we could leave our discussion of animal signalling here with the conclusion that most signals are probably revealing handicaps anyway. However, the confusion in the literature that has already arisen as the result of people failing to recognize the implications of this conclusion has now become so great that we cannot honourably take this course. Faint-hearted readers should skip the next paragraph and go straight to the next section. What follows next is enough to make anyone throw up their hands in despair.

As we have seen, Grafen showed that Zahavi's insight into the nature of animal signalling could lead to evolutionary stability and so could work in theory. Strategic choice or conditional handicaps are therefore viable. Many people have then assumed that honesty in animal signalling is only achieved through the signals being handicaps in this sense. However, as we have also seen, revealing handicaps also provide a way in which signals can be honest and, given the evidence, many signals may well turn out to be revealing handicap signals. Unfortunately, Grafen has taken the step of saying that revealing handicaps should not be called signals at all on the grounds that only strategic choice handicaps are true signals. ('The revealing handicap does not operate as a signal because the content of the message is directly observed', 1990a, p. 538).

This is surely the ultimate tyranny of definition over reality. Since we do not know whether there are any strategic choice handicap signals in practice, it seems absurd to say that only they can be counted as 'signals' and that other obvious cases that everyone else would call signals without even thinking twice (such as roaring in red deer) should not be because they may be 'revealing' handicaps. We need to go back to an earlier definition and stick with the 'specially evolved' criterion that covers all kinds of signals because for all of them there hangs over their heads a much more important question than what they should be called. This is whether the selection pressure on receivers to respond only to 'true' information has resulted in signallers always being completely truthful.

Are all signals 'honest'?

As we have seen, Dawkins and Krebs pointed out that the constant danger of being manipulated by signals conveying false information that arises whenever the interests of two animals do not coincide results in evolution towards cheat-proof signals. Then we saw that cheat-proof or 'honest' signals carrying 'true' information can arise either through being conditional handicaps or being revealing handicaps. But is the selection pressure not to be manipulated on the part of the receiver sufficiently strong to guarantee that all signals end up being honest?

The widespread existence of mimicry makes it difficult to believe so. Batesian mimics are clearly transmitting false information about their palatability and derive their protection from the fact that most of their predators have learnt to associate certain colour patterns with distasteful prey and leave them alone. Overall, the predators gain from this behaviour because most of the brightly coloured potential prey items they come across will be genuinely distasteful. The predators gain from not being stung or poisoned even though on those occasions when they are fooled by a palatable prey's resemblance to a poisonous one, they could, in that instance, be said to be 'manipulated' by the mimic.

Batesian mimicry is, of course, an example of false information passing from one species to another, but within-species mimicry may also be commoner than we had thought. For example, individuals that are not particularly strong (or even, as in the case of the mantis shrimp, have temporarily lost the use of their claws as weapons through having moulted) may mimic genuinely strong or dominant members of their own species and 'get away with it' on a small number of occasions (Adams & Caldwell, 1990; Wiley, 1993). Maynard Smith & Harper (1988) argued that the status badges of some birds (where patches of feathers seem to signal dominance) could in some instances be used by cheats to signal false information about how aggressive they were likely to be if challenged. Birds sporting 'high status' badges pay a cost of increased numbers of challenges from other dominant birds (Møller, 1987) but this may not be sufficient to eliminate cheats altogether, particularly where the value of the resource being fought over is low compared to the cost of challenging. 'Strategic choice' handicaps will only result in signallers that are honest on average. Some signallers could give out false information and get away with it sometimes provided that the costs of being found out are not too great.

For a somewhat different reason, too, both strategic choice and revealing handicaps may be less than 'honest'. It may simply be that there is no signal that could give the information required – that is, there may not be the raw material for 'true' information to be signalled. Suppose you were confronted with a field of 20 marathon runners and you had to devise a test to 'reveal' who was going to win before the race took place. You could use any test you wanted to – getting everybody to run short distances, lifting weights for hours on end – anything except actually getting them to run the full 26 miles itself. It could well be that you would fail to come up with anything reliable. There just might not be any test that you could apply, and successfully predict the outcome, however costly, however full of effort. Similarly with animal communication. Some signals such as roaring in red deer or tail beating in cichlid fish might be reasonably closely correlated with fighting ability or whatever quality was being signalled about, but none of them perfectly so. Body size and associated signals of colour or pattern that accentuate size, might be enough to separate the reasonably good fighters from the no-hopers, but perhaps no signal could be found that would signal absolutely guaranteed 'true' information about what would happen in a real fight.

So the answer to the question 'are animal signals honest?' has to be strongly qualified. There will certainly be selection pressures on signals to make them bearers of 'true' information but a few less-than-honest signals could creep in even in the best ordered strategic choice handicap systems. And then, with both revealing and conditional handicap, there may not be any possible way of signalling cast-iron guaranteed 'true' information because there may not be any way of gauging the quality in question with complete reliability. By all means we can call animal signals 'honest' but we should only do so if we are prepared to settle for the fact that they are not going to be 100 per cent honest 100 per cent of the time.

CONCLUSIONS: INFORMATION TRANSFER OR MANIPULATION?

Despite the difficulties that surround the definition of 'communication' and 'signal', it is clear that there have been strong selection pressures for animals to influence and, in some cases, be influenced by, the behaviour of others. Many signals have become 'specially evolved' for this function as was recognized many years ago by pioneering ethologists such as Tinbergen. Animals can be said to 'transfer information' during communication and many aspects of signal design, such as conspicuousness, can be seen as aids to information transfer. There are other signals (called 'conspiratorial whispers' by Dawkins & Krebs, 1978) which are not particularly conspicuous or obvious in any way but which function to transfer information between members of flocks, schools or herds quite effectively because the animals are close enough together that they can see or hear each other without any special adaptations.

It is also clear, however, that in addition to the selective pressure to transfer information effectively, there is often a second pressure to manipulate or deceive other animals and a consequent counter-selection on those animals not to be deceived. Many signals have evolved to be reliable and truthful, either through being strategic choice 'handicaps' or through being closely correlated with the quality being signalled about (that is, revealing handicaps). Receivers have, in turn, been selected to repond to honest rather than dishonest signals.

In recent years, the quest for 'honest' signals has dominated the study of animal communication, with information transfer being relegated to at best a minor role. However, it should be clear by now that while selection for signal honesty has been one important factor in the evolution of animal signals, it has not been the only one. Honesty is only the first step. Getting the honest (or, as we have seen, sometimes not so honest) message over to the receiver is the next. In other words, the selection pressure for effective information transfer that was stressed by the early ethologists is still going to be important. Furthermore, we now know considerably more than they did about how effective information transfer can be achieved. We know, for example, more about the physical properties of signals such as their sound frequencies (Romer, 1993) and colours (Lythgoe, 1979; Endler, 1993) that enable them to travel effectively through the medium separating signaller and receiver. We know more about the 'fit' between signal and brain and the sense organs of the receiver (Guilford & Dawkins, 1991). It is now apparent that effective information transfer can mean all sorts of tricks and devices by which signallers influence or attempt to influence the behaviour of receivers. Male jumping spiders use their legs to stimulate the prey-catching response of a female to get her attention in the initial stages of courtship (Clark & Uetz, 1992) and some male cichlid fish have egg-like spots on their anal fins that the female attempts to take in her mouth as if they were real eggs and so stimulates the male to release sperm (Wickler, 1962; Hert, 1989).

So there not just one but two selection pressures operating on signals – one for effective information transfer and one for manipulation (on the part of the sender) plus resistance to manipulation (on the part of the receiver). To say that animal communication must be one or the other is therefore quite misleading and to concentrate on one selection pressure to the exclusion of the other leads to some very incomplete views of animal signals. Nowhere is this one-sidedness more apparent than in recent studies of sexual signals.

Further reading

The volume on *Communication* edited by Halliday & Slater (1983) gives a very comprehensive introduction. A special issue of *Philosophical Transactions of the Royal Society* (**340**: 161–255, 1993) provides a more up-to-date coverage. Guilford & Dawkins (1993) stress the role of receiver psychology in signals, a subject that is covered more fully in the next chapter.

6

SEX AND SEXUAL SELECTION

Male bluehead wrasse (*Thalassoma bifasciatum*). Photograph by Marian Stamp Dawkins.

Sex is a surprisingly difficult and misunderstood subject. There are things about it that nobody understands, such as why it is so common as a method of reproduction and why some male animals adopt sexual ornaments that are so elaborate and conspicuous that they endanger the males' own survival. As we will now see, the evolution of sex and the evolution of sexual signals are intimately bound up together and both are shrouded in misconception and mystery. Both the mistakes and the mysteries will form the basis of this chapter.

THE ADVANTAGES OF SEX

It has been a fundamental idea in this book that genes are passed from one generation to the next in proportions that are affected by the behaviour of the animals carrying them. Feeding, escaping from predators or building nests are all activities that help in this process. All, that is, until we look at the final generative act. When animals get to the point of making a new generation, a most extraordinary and counterintuitive thing happens.

Instead of the genes that have served them well enough in their own lives all being passed into an offspring, only half of them go. The rest of the offspring's genes are *from another animal.* The offspring are thus semi-aliens, not flesh of their parent's flesh but only half so.

From the genes' point of view, the situation is even odder. Engaged, so we had thought, in a bitter struggle for a place in the gene pool, they appear to step aside at the last minute, most of them accepting a 50 : 50 chance of entering the next generation instead of the certainty they could have had with asexual reproduction in which all the genes are passed on. And yet sexual reproduction persists! It is a very widespread method of propagation. The mystery is what possible advantage there could be to it.

Early explanations, that it was 'good for the species' to have the variation that is brought about by the mixing and shuffling of sexual reproduction, had largely to be discarded. It was an attractive idea to think that sex persists to enable a population to 'cope' with some future environmental change even though individual animals might do better to reproduce asexually and so avoid the cost of sex. But by the end of the 1960s, the weakness of such arguments was generally realized. A trait – particularly one as costly as sexual reproduction – would not persist if it was detrimental in the short term to its bearers, however beneficial it might be in the longer term. We cannot expect a trait, to use Sydney Brenner's immortal words, to evolve in the Cambrian 'because it might come in handy in the Cretaceous'. Something other than being 'good for the future of the species' must be invoked to account for the persistence and widespread occurrence of sex.

The problem was that nobody could think of anything very convincing. There were some attempts to argue that producing offspring that are different from their parents would benefit an individual parent because the environment might change quickly from one generation to the next, but in reality environments did not seem to change that much, at least not drastically enough to compensate for the apparently massive disadvantage to sex in a constant environment.

Williams (1975) suggested that environments might effectively be thought to be changing if offspring dispersed widely and so ended up in very different places from their parents. It then might be advantageous for the parent to produce many different sorts of offspring, enabling at least some to survive in their new habitats, particularly if several offspring landed in one place and competed with each other. 'Sib competition' might favour sex. Bell (1982) took the sib competition idea even further, but based it not so much on adaptation to an environment changing over time but to an environment varying over space. He argued that the diverse progeny of a sexual brood would occupy slightly different niches from one another and so compete with one another less fiercely than the uniform progeny of an asexual brood, all of which would have exactly similar ecological requirements. More sexual than asexual offspring might therefore result.

Then, during the 1970s, it was realized that environments might not be as constant as they seemed. The physical aspects of an environment – temperature, rainfall, amount of vegetation and so on – might not change very much, but the environment made up of other animals could be changing constantly. Competitors, prey, parasites and predators would all be evolving their own adaptations and counter-adaptations and presenting constantly changing selection pressures that might provide the driving force for sexual reproduction.

Hamilton (1980) argued that the only one of these pressures that was strong enough to overcome the disadvantages of sex was parasites. A host genotype that is very successful in resisting the commonest parasites now will be very much less successful in the future because by that time the parasites will themselves have shifted their own genotype frequencies and the host will no longer be resistant to the commonest parasite type. The only way in which an animal could equip its descendants to deal with the parasites that will be around when they grow up is, Hamilton argued, to reproduce sexually with all the possibilities that gives for rapid change. Certainly an animal's predators, prey and competitors may also be evolving at the same time and providing some change to its environment but the massive asymmetry in potential speed of evolutionary change between parasite and host (the fact that disease organisms can evolve so much faster because of their shorter generation times) means that it will be parasites that have the overwhelming effect. We have only to think of the rapidity with which the malarial parasite evolves resistance to the various drugs that have initially worked against it to realize how quickly disease organisms can change and therefore what any slow reproducing animal is up against. Parasites cause constant evolutionary time lags. Hosts are always 'out of date'. The only way in which they can even stand a chance of catching up is to pay the cost of sex in return for the variation that it brings and the possibility that at least one of their offspring will hit upon a gene combination that confers resistance.

In 1862, Charles Darwin wrote of the problem of why animals reproduce sexually as opposed to asexually: 'The whole subject is as yet hidden in darkness.' In 1990 Hamilton *et al.* wrote: 'Darwinian theory has yet to explain adequately the fact of sex' (p. 3566). Increasing numbers of scientists believe that Hamilton's own theory of parasites and sex provides one of the few convincing explanations of a subject that remains genuinely problematical.

Selection on males and females

Whatever the advantage of sexual reproduction, males and females do exist and their behaviour is often very different. Males, with their large numbers of small gametes, are potentially capable of fertilizing the eggs of many females with their larger, but restricted number of eggs. But, of course, gamete size is not the only difference between the sexes. Developing embryos are often nurtured and protected through a long period of infancy and usually one parent gives more of this care than the other. Although there are cases where the male alone cares for the offspring or where male and female share parental duties more or less equally, it is more commonly the female that is the one investing most in each offspring and taking more 'time off' to rear young than the male. Females are thus generally more limited in their reproductive possibilities than males and at times this difference can reach dramatic proportions. In polygamous species such as elephant seals and sage grouse, the female does all the nurturing and caring for a relatively small number of offspring while a few successful males may mate with many females and father a very large number of offspring, contributing little to each one except the genetic material contained in the sperm. At this point, a fallacy potentially obscures the account of male and female behaviour.

The exploitation fallacy

It is very easy to assume from all this that males 'exploit' females, 'getting away with' large numbers of offspring at virtually no cost to themselves. From this, it is but a short step to believing that in some sense it is 'better' to be male than to be female, at least in species where the male does not help with the care of the young. What this fallacious line of reasoning ignores is that males face many other costs besides the admittedly minimal one of the single sperm that manages to fertilize the egg. To achieve this apparently cheap result, the male will have had to make millions of other sperm which are simply wasted. Then, in order to achieve this mating, the male will have had to fight to gain the female or to defend a territory to which a female will be attracted. Fighting is often such a costly business that even the most successful polygamous males enjoy only a brief reign. The harem masters among elephant seals and red deer have very much shorter reproductive lifespans than the females they, temporarily, defend (Clutton-Brock *et al.*, 1982).

Females may thus invest more in individual offspring than males, but the males pay a cost in other ways. In fact, the greater the reproductive potential of a single male (the lower his parental investment), the higher will be the toll extracted from him in fighting and in competition with other males because they too will be battling to fulfil their reproductive potentials. The less each invests in parental care, the more of them there will be around to fight over females.

An individual male may be reproductively more successful than any female, but for every greater than averagely successful male, there will be other males that are less successful than average. The *average* success of males, as a whole, will be exactly equal to the *average* success of females. If it were not, the sex ratio would shift over evolutionary time towards increasing the frequency of the more successful sex. In a population of females, parents producing males would have an overwhelming advantage and vice versa (Fisher, 1958). The population sex ratio stabilizes when the average success of each sex is exactly equal. (This does not necessarily mean that the *numbers* of males and females will be exactly equal; parents may invest more in individual offspring of one sex, say males, if by so doing they increase that offspring's chances of being reproductively successful. However, if they give more to individual offspring of one sex, they make fewer of them, so that what is equalized is the total *investment* in the two sexes.)

Males cannot therefore be said to be more successful than females or, as a sex, to exploit them, if 'exploitation' means that the females are at a disadvantage. Females have, on average, just as many offspring as do males. An individual male may have a more successful reproductive career than any one female, but even in a polygamous species, on average the hard-working females are doing just as well as the more flamboyant or aggressive males, some of whom will have no reproductive success at all.

Because males in such species have both a much higher reproductive potential than females and also a much higher chance that, in practice, they will get no matings at all, there may well be severe competition between males. The females can get on with the day-today business of surviving and rearing offspring, but the males, as Darwin (1871) realized, may evolve characters that threaten their very survival simply because they give an advantage in fighting other males or attracting females. Bright colours, long tails, plumes, antlers and horns fall into this category. If male birds evolve long tails that attract the attention of predators and hinder their flight but do so because it attracts females, it is difficult to decide which sex is exploiting which. It is much better to drop the idea of 'exploitation' altogether and concentrate instead on the different strategies that the two sexes can adopt, because even without any notions of one sex doing better than the other, we find confusion enough to keep us going for the rest of the chapter.

Male rivalry and female choice

Darwin recognized that certain males might have an advantage in the competition over mates if they possessed characters that either helped them to win fights with other males or to attract females. Characters that gave males the edge in either of these two ways, Darwin referred to as 'sexually selected'. He specifically excluded from sexual selection any characters that may be useful in reproduction but were not selected by this competition with other males. Characteristics that helped a male to rear young, for instance, might have an effect on reproductive success, but unless it helped him to obtain a mate in competition with other males, Darwin did not refer to it as evolving under sexual selection.

Darwin did, however, make a distinction between characters that evolved because of intra-sexual competition (male–male competition) and those that evolved through inter-sexual competition (female choice). Some characters seem to fall conveniently into one or other category – horns and antlers, for example, seem to be weapons directed at other males whereas plumes and crests appear to be directed at attracting females. But there are many other cases when it is very difficult to make a clear distinction and where, indeed, it may not be very fruitful even to try to do so. Many female animals, from guppies (Kodric-Brown, 1992) to mice (Coopersmith & Lenington, 1991), have a preference for dominant males, that is, males that can hold their own against other males. Female elephant seals have an even more blatant way of 'using' male fighting ability. When mounted by one male, the female calls in a way that attracts the attention of other nearby males. If one of them is bigger and stronger than the male that is mating with her at the time, that male will oust her current mate and copulate with her himself (Le Boeuf, 1972). The female's behaviour thus exerts a major influence on which male fathers her offspring but she does not 'choose' her mate in any conventional sense. Rather, her behaviour stimulates male–male competition, leaving her with the strongest and most dominant male in the vicinity.

Male–male competition and female choice are often, therefore, inextricably bound up together and from the female's point of view it is easy to see why this should be so. It will clearly benefit her to choose as a mate a male that is strong, healthy and free from debilitating parasites. What better way to do that than to allow males to fight, or even provoke them into doing so, and then choose the winner? A male that can withstand aggression from other males and hold his place in a dominance hierarchy must be among the better quality males that are available and by using interactions between males as her guide, the female is assured of a high quality mate at little cost to herself.

Using male–male interactions in this way seems such an adaptive strategy for females that we might ask why females do not always use it and, in particular, why they should ever be attracted to red wattles or long tails, which would seem to have a much more tenuous link to male 'quality'. But is there possibly a closer connection than appears at first sight? Borgia (1985) describes the extraordinary bowers built by the male satin bowerbird: two walls of interwoven sticks with a 'stage' at one end on which the male places anything blue he can find. Blue flowers or blue feathers or even blue plastic will do. Since the males are constantly fighting and raiding each other's bowers, a well-kept bower with many blue objects is effectively an indication of the owner's fighting ability. A female that is attracted to such an ornamental 'folly' is therefore choosing a strong and healthy mate even though she may not have seen him interacting with other males.

It is possible – although very difficult to demonstrate one way or another – that many other male ornaments such as plumes or patches of iridescent feathers have a similar link to male health and vigour through their role in male–male interaction, with females being attracted to them as signs that the male has passed some sort of test with other males. Many bird species signal dominance through patches of feathers – so-called 'badges of status' (Maynard Smith & Harper, 1988), such as black throat patches in house sparrows (Møller, 1987). Although potentially easy to fake (a subordinate bird could seemingly grow the relevant feathers and so give a dishonest signal), there are enough checks that the badge gives an overall honest indication of male quality. Birds sporting high status badges receive more aggressive challenges from other dominant birds (Moller, 1987) and so pay some costs not paid by birds in subordinate plumage. A male bird with dominant plumage in good condition is therefore, like the bowerbird with his well-equipped bower, demonstrating that he can stand up to other males with impunity. By being attracted to the status badge, a female gains a fit and healthy mate. What looks like 'pure' female choice could thus in many cases be female choice linked to male quality via male–male interaction.

For this reason, it is important not to see inter- and intra-sexual selection as two separate processes to be prised apart at all costs. Male fighting ability is such an outstandingly good – and honest – indication of those qualities that will increase offspring survival (health, disease resistance, strength) that it would be strange if females did not make use of it in some way. And, given that males do not spend all their time fighting and often settle disputes with signals that, as we saw in the last chapter, will be selected to be 'true' indications of their fighting ability, we would expect females to be attracted to these 'true' signals as well.

This has important evolutionary consequences. If a signal of male dominance also has the effect of attracting females, this double advantage to the male may do the oddest things to the signals he gives. Sexually selected signals are among the most striking and problematical of all aspects of animal communication.

THEORIES OF FEMALE CHOICE

Sometimes females choose males on the basis of material possessions such as food or good nesting sites (e.g. Thornhill, 1976; Alatalo *et al.*, 1986). Here the male is essentially a provider or protector, giving the female and her offspring immediate and direct benefits. If his abilities as provider and protector are heritable, the female will gain additional genetic benefits from mating with him because she will have sons that also have these traits but even if without this added benefit, the female is better off mating with him than with other less well endowed males. In some species, however, the males provide no such immediate benefit and the females appear to be attracted to some attribute of the males themselves. The controversy that Darwin (1871) provoked by his explanation of this is still with us today.

We have already seen that Darwin proposed that male ornaments such as long bright tail feathers in certain bird species evolved to attract females even though they might be positively detrimental to the survival of the males themselves. Central to Darwin's theory was the idea of female choice: what females were attracted to would determine how the male ornament would evolve. This element of sexual selection led many people to be sceptical about the theory itself on the grounds that female fancy seemed too whimsical a thing to explain the evolution of traits that could actually threaten male survival. In this, as in so many other instances, Darwin's understanding was more profound than that of his critics. Sexual selection was not some extra force in opposition to natural selection. Rather, natural selection included successful reproduction and if that involved attracting a lot of mates and having a short life rather than a long life and no matings, then a brief but glorious reproductive career would be selected for.

To date, there are at least three major theories of female choice, each one beset with difficulties and each hampered by confused thinking. The first theory, proposed by R. A. Fisher (1958) and now known as the 'Runaway Theory', is a direct descendant of Darwin's original idea. Fisher argued that if a majority of females had a trait of preferring a particular sort of male, however bizarre his appearance, then other females would be favoured if they chose that same sort of male to mate with because they would have sons that would be attractive to many females. Females choosing in the same way as the majority of other females would 'cash in' on the male's success through having sons like him who would, in turn, be attractive to lots of females.

There is thus a built-in positive feedback in the system. The more common a particular kind of male becomes, the more advantage accrues both to the males of that type (because that is what females choose) and to the females preferring them (because they will have sons that other females will choose). The system 'runs away', until all males have this trait. There is another sense in which the system 'runs away', too (Bateson, 1983). If the females are following some simple rule such as 'choose the male with the longest tail', tail length will increase over evolutionary time as males with longer and longer tails gain more and more of the matings. Tail length will 'run away' until finally halted by the influence of ordinary natural selection. If a male's tail becomes so long and colourful that he is eaten by a predator before being able to reach any females, he will obviously be selected against, despite the enthusiasm with which he would be greeted by females if he were able to survive that long.

Lande (1982) and Kirkpatrick (1982) constructed genetic models of Fisher's originally verbal argument and showed that Fisher was right in the sense that this sort of sexual selection does produce both sorts of 'runaway' effect. The genes for producing exaggerated ornaments in males and those for preferring such ornaments in females will come to be found in the same bodies because these will be the offspring of fathers with the adornment and mothers with a preference for that adornment. Whatever the success of the offspring, both sets of genes will be favoured, one through being expressed, the other being carried along by the success of the set that is expressed. Each feeds on the other's success and the two together sweep through the population. Fisher's theory shows how, despite the threat to individual survival that male adornments may pose, such characters could evolve. But it is not without its critics.

One objection to Fisher's theory is that it only works once a majority of females in a population have a preference for a particular male ornament. And why should there be such a majority with such an odd preference? It is not, in fact, difficult to see how it could happen. Perhaps slightly longer than average tails initially helped females to discriminate males of their own species from those of other species; perhaps they helped males to fly a bit better; perhaps they were just more conspicuous from a bigger distance. There are many reasons why long tails could be favoured even when their bearers are in a minority and why long tails could increase within the population until they reach the critical 'take-off' frequency for the runaway process to occur.

Another objection is more serious. It is clear from Lande and Kirkpatrick's models that the runaway process will eventually be halted by natural selection. Male tail length will not go on increasing indefinitely, but will stabilize at a balance point between the advantage of being attractive to many females and the disadvantage of having such a cumbersome and conspicuous ornament. Eventually all the males in the population will end up with a tail length at this optimum compromise point. Parker (1982) then pointed out that this would completely negate all the advantages of female choice. If all the males were genetically the same, there would be no genetic advantage to a female in choosing one rather than another. And if there were any *cost* to female choice (such as having to travel around looking at different males), then unchoosy females that paid no costs would be favoured over choosy ones that paid costs and got nothing for it. From a genetic point of view, females might just as well mate with the first male of their own species they came across regardless of whether he were ornamented or not.

It is possible to give a partial answer to this objection to Fisher's theory (Parker, 1982). The bright, gaudy displays that are predicted by Fisher's theory actually have the effect of *reducing* the costs of female choice because females can see or hear males possessing them from a greater distance and so do not have to travel so far between mates to make their choices. However, this may not reduce the costs of female choice sufficiently to make Fisher's explanation viable (Pomiankowski *et al.*, 1991). This is one reason why a serious rival to Fisher's theory has recently gained popularity as an explanation of female choice. This theory (or more accurately group of theories) has rather unfortunately come to be known as the 'good genes' theory.

The reason that this name is unfortunate is that, by contrasting it with Fisher's explanation, there is an implication that Fisher's theory is not a theory of good genes, when of course it depends critically on the idea that sons inherit the genes that gave their fathers their attractive qualities. 'Good' genes for Fisher's theory means genes for attractive qualities in sons, whereas 'good' for the 'good genes' theory means genes that confer benefits such as health, strength and vigour on both daughters and sons. Since the terminology is so deeply entrenched in the current literature, however, the term 'good genes' will from now on be used in the second, non-Fisherian, sense, confusing though that is in some ways.

It was Amotz Zahavi who spearheaded the idea that female choice must be based on something more substantial than the female 'whim' that formed the essence of the Darwin–Fisher explanation of sexual ornaments. Zahavi (1977) argued that that more substantial something was the male's health and vigour which the female could gauge through looking at his adornments and displays. A male with a 'handicap' (see Chapter 5) such as a long tail would effectively demonstrate that he had great physical prowess in order to survive despite the handicap and so a female which mated with such a male would ensure that she had strong healthy offspring. The changing fortunes and nuances of Zahavi's ideas were discussed at length in the last chapter and so we will not go into them again here except in so far as to remind ourselves that there are two distinct kinds of handicaps (Maynard Smith, 1991).

With conditional or strategic choice (Grafen, 1990a, b) handicaps, the growing of an ornament such as a long tail is costly to a male and, furthermore, it is costlier to males in poor condition than those in good condition and so constitutes a 'handicap' (it reduces fitness). Females choose males on the basis of their body condition, using the size of the ornament as an indication of the quality or condition of the male. Assuming that strong healthy males tend to sire strong healthy children, the female ensures that her offspring receive 'good genes' from their father by choosing a male with a big handicap.

With revealing handicaps, there is no necessary reduction in fitness as a result of giving the signal. In the context of sexual selection the most widely discussed version of this was put forward by Hamilton and Zuk in 1982. They argued that the most important 'good genes' that a female could obtain for her offspring would be genes for resistance to parasites and diseases. This idea grew out of Hamilton's (1980) conviction that sex itself owed its maintenance to the potential it gave to sexually reproducing animals to come up with new varieties in the never-ending evolutionary battle against disease organisms. And if the answer to the riddle of sex was parasites, it was for Hamilton the obvious next step to say that it was also the answer to the riddle of why females should be attracted to elaborate male ornaments: they enabled females to pick out disease-free males.

Hamilton did not, however, say that the ornament lowered the male's resistance to disease and so constituted a cost that was paid in terms of reduced physical health, as would be implied by the conditional or strategic choice handicap theory. Rather, he said that the ornaments simply *revealed* state of health without damaging it and so constitute revealing handicaps, a bit like a survey done on a house. Surveying a house does not make the fabric of the house worse – it simply shows up the defects that are or are not there. The more extensive the survey, the more confidence can be placed on the conclusion that the house is structurally sound. Long tails, inflated throat pouches or iridescent plumage, Hamilton and Zuk argued, enabled females to make very extensive surveys of males since external parasites would show up directly on patches of skin or bright feathers and internal parasites would make their presence felt by indirectly affecting the state of these outward and visible signs.

It is important to realize that both these versions of the 'good genes' theories – the strategic choice handicap and the revealing signal idea – are direct alternatives to the Darwin/Fisher view of things. On Fisher's theory, female choice is based on whim – the elaborate male ornaments are attractive in themselves. They signal nothing else about the male than that he possesses them. As long as most other females have the same whim, the adornment will persist and evolve because the females that are attracted to them will gain through having attractive sons.

On the 'good genes' theories, however, female choice has more substance. The males are assessed for their health and disease-free vigour and the assessment cue that the females are using is no mere arbitrary whim. It is related in a cheat-proof way to an ornament that can be sustained or flaunted only by genuinely healthy males in good condition. By choosing such a male, a female gains strong healthy children of both sexes.

It is equally important to realize, however, that the two versions of the 'good genes' theory – the conditional choice handicap and the revealing signals – are also very different from each other. A failure to realize that there are two distinct versions has led to no end of confusion. For example, Balmford *et al.* (1993) argued that it is possible to distinguish between Fisherian and good-gene theories of sexual selection by looking at the aerodynamic costs a male bird incurs as a result of growing a long tail as a sexually selected ornament. They argued that forked tails are unlikely to have evolved as honest signals of male quality because, at least in the initial stages of evolution, longer feathers actually improve flight efficiency. They are therefore most likely to be Fisherian traits. On the other hand, what they call graduated tails (ones where all the feathers are elongated except for the outer ones) are likely to be indicative of 'good genes' because they do impose a cost in flight efficiency. These are therefore most likely to be handicaps.

By assuming that 'good genes' are always indicated by costly handicaps, Balmford *et al.* make the mistake of thinking that the only alternative to Fisher's theory is a strategic choice handicap. They quite fail to realize that 'good genes' can just as reliably be indicated by revealing signals that may not be particularly costly to give. Forked tails that do not interfere with flight but simply 'reveal' health or size could thus be indicating 'good genes' just as well as tails that impose a flight cost on their bearers. And they, like so many other workers in this field, also do not consider the third explanation of female choice and male ornamentation, the one that we have not looked at so far but one that in many ways offers both the simplest and most convincing explanation of any of the theories so far proposed.

This third explanation could be said to be quite literally staring us in the face. Loud sounds, colourful plumage and ornaments such as long tails (especially when moved around) are, quite simply, more conspicuous and more likely to attract the attention of a female from a bigger distance than less intense signals. The evolution of exaggerated sexual signals may, therefore, have nothing to do with demonstrating male 'quality' or showing off good genes in a handicap way at all. They may just be the result of selection on males to be noticed by females ('passive attraction' Parker, 1982). Arak (1988), for example, has shown that, instead of 'choosing' one male rather than another, female natterjack toads simply move towards the louder of two sounds. This would obviously put selection pressure on males to croak loudly so that they increase their chances of the females moving towards them.

There are good reasons for thinking that this is not an isolated phenomenon. A well-known finding by early ethologists is that many animals respond to 'supernormal' stimuli – that is, to stimuli that extend or exaggerate a naturally occurring stimulus. Tinbergen (1951) described how oystercatchers would try to incubate ostrich eggs rather than their own – a case of the bigger the better. Magnus (1958) showed that male silver-washed fritillary butterflies would prefer to court a revolving drum that flashed orange spots at a rate of 75 'wing beats' per second to a real female of their own species that could only manage eight wing beats per second. Here faster is better.

If response to supernormal stimuli is widespread, then it is easy to see how 'exaggerated' (bigger, faster, louder, more colourful) ornaments would evolve, with no necessary involvement of 'good genes' at all, in either a Fisherian or a handicap sense. Females that chose males with conspicuous, easy to recognize signals denoting which species they belonged to would mate successfully with little cost. Such 'easy choice' behaviour would be favoured, even without the female gaining an additional advantage through the attractiveness of her sons (Fisher's theory) or genetically high quality children of both sexes (handicap theory).

Of course, she would benefit even more if she did gain in either of these two additional ways, but the point is that even if she did not, we would still expect selection for conspicuous signals. Choosing a male of the right species and one which is close and ready to mate is of such an advantage in its own right that we could expect females making such a choice to be strongly selected for and in turn to select for males that were conspicuous enough for females to reduce their choice costs. Given the tendency of many animals – particularly birds – to respond to supernormal stimuli, it is easy to see how males could evolve so that they had exaggerated versions of these 'easy choice' signals. Both sexes would benefit, the males because they would be most likely to be found first, the females because they would find a mate with minimal cost.

Ryan & Rand (1993) have suggested that this third theory of sexual selection should be called 'sensory exploitation' to emphasize the fact that males evolve signals to 'exploit' the ways in which the brains and sense organs of females actually operate. They give the example of the curious mate preferences in a group of closely related frogs. In one species, the tungara frog, *Physalaemus pustulosus*, the males produce calls that are described as a 'whine' followed by between zero and six 'chucks'. Female tungara frogs prefer males that have large numbers of 'chucks' to their calls. In another species, *P. coloradorum,* males have a similar 'whine' but no 'chucks' to their mating call. The odd thing is that *P. coloradorum* females are actually more attracted to artificial calls with some *P. pustulosus* '-chucks' added to the 'whine' of their own species than they are to the pure whine of their own species which does not naturally have any 'chucks' at all. They argue that female *P. coloradorum* have a pre-existing 'sensory bias' inherited from the common ancestor of both species which has been uncovered by their experiment. Male *P. pustulosus* frogs, on the other hand, have already responded to this sensory bias naturally and evolved calls that contain 'chucks' that are, because of the make-up of the female's auditory system, very attractive to females. In this sense the males exploit the pre-existing female preference.

While this is an important example because it shows that male signal and female response can be related to each other in unexpected ways, there are two reasons why the term 'sensory exploitation' may not be the best one to describe this third theory of female choice. First, we should be wary by now of the implication that any animal regularly 'exploits' or 'manipulates' another unless it is to the overall benefit of the one that is 'exploited' to do so (thereby removing both the sting of the term 'exploitation' and the point of using such a word at all). It is much better to stick to some neutral term such as 'sensory bias', 'efficacy' or even just 'effectiveness of information transfer' (Guilford & Dawkins, 1991) and to begin to explore the ways in which sexual ornamentation could be explained as simply the best way for males to stimulate the brains and sense organs of females. This would enable us to relate the evolution of sexual signals to the more general principles of communication we were considering in the last chapter. There we saw that selection on signaller to give effective signals goes hand in hand with selection on receivers to respond only if it is to their advantage to do so. Females are thus not exploited (to their detriment), they benefit.

Second, 'sensory bias' focuses on one small aspect of what makes a signal effective, namely a peculiarity of sound perception in one group of frogs. What is of much more interest are the *general* properties of signals that make them into effective carriers of information. As we saw in the last chapter, these include characteristics such as sound frequency or colour that transmit effectively through particular environments and features such as suddeness of onset or contrast with the background that make them conspicuous to the eye or ear of the receiver. What is needed is a term that will cover both the specific sensory biases that may exist in some species and the more general properties of signals that make them effective at stimulating receivers. The classical word 'ritualization' might do, but its old-fashioned ring is perhaps unfortunate. 'Sensory bias', seen as part of the more general selection on signals to be effective at getting their message across (efficacy of information transfer), is therefore better.

Conclusions

There are, as yet, no definite answers to the question of why sex evolved or which of the three current theories of female choice is right. As we have seen, there are objections to Fisher's ingenious theory of the evolution of sexual ornamentation. Equally, critical evidence in support of a 'good genes' theory of female choice is still lacking, despite the considerable attention it has attracted. As with other areas of animal communication, so much emphasis has been placed on the idea of signals as honest indications of 'quality' (such as health) that alternatives, such as the sensory bias theory, have been given little consideration.

In order to see how good an explanation a 'sensory bias' theory would be of male ornamentation and female choice (or indeed of any signal) and in particular how well it fares in comparison with its main rivals the Fisher theory and the good genes theories, we would obviously need to know quite a bit more than we do now about how the brains and sense organs of female animals (or indeed any animal that is on the receiving end of a signal) actually work. That brings us to a critical stage in this book.

So far, we have considered mainly evolutionary or adaptive questions about animal behaviour. The confusions we have tried to unravel have mainly been misunderstandings of the way in which natural selection operates on behaviour and to a large extent we have treated animals almost as if they were nothing but disembodied carriers of genes. The issues we have been concerned with have been those of how and why some genes get into the next generation and others do not. What those carriers are, what they are made of and how their bodies work have been largely pushed to one side, as if one could discuss evolutionary theories and ignore the flesh and blood animals that are the result of evolution. Now, with our discussions of communication and sexual selection, we have seen the limitations of such an approach. Evolutionary speculation is all very well and can, as we have seen, lead to some fine and exciting theories. But there are dangers in thinking of animals just as hypothetical entities that can do anything and go anywhere. How their sense organs receive incoming information, how their brains process and store it, how their actions are limited by what their nerves, muscles and storage systems can do are not only interesting issues in their own right. They also affect what evolutionary theories are likely and even plausible.

We are not, of course, dealing with an imaginary set of gene carriers that might, hypothetically, have evolved in any direction, but with the actual set of real animals that follow rules such as 'respond to the largest egg (longest tail, loudest voice) around' or have memories that can only hold limited amounts of information or eyes that can only detect light of certain wavelengths. To know why animals behave as they do, we have to know both what constraints their bodies have imposed on them and what opportunities have been opened up to them. We have, in other words, to know something about the machinery of behaviour.

Further reading

Maynard Smith (1991) treads a clear path through the maze of sexual selection theories while Gould & Gould (1989) provide useful examples of different sorts of sexual behaviour. Clutton-Brock (1991) describes different patterns of parental investment.

7

The machinery of behaviour

Mute swan (*Cygnus columbianus*). Photograph by Mike Amphlett.

There was a time when understanding how animal bodies worked was considered just as important to the study of animal behaviour as understanding evolutionary adaptations. In 1963 Niko Tinbergen published a famous paper called 'On aims and methods in ethology' in which he said that there were four kinds of questions that could be asked about animal behaviour: 'How does it work?', 'How does it develop in the lifetime of the individual?', 'What were its evolutionary origins?' and 'What is its survival value?' Tinbergen himself believed strongly that none of these questions should be asked in isolation and so in his study of eggshell removal in the black-headed gull, for instance, he would constantly move backwards and forwards between questions about adaptation ('Why is it adaptive for a gull to remove eggshells from its nest?') and those about mechanism ('How does a gull recognize an egg shell?').

Since that time, however, there has been an increasing tendency, at least until very recently, for people studying animal behaviour to split into two groups, roughly speaking those interested in mechanism and those interested in adaptation. Those interested in mechanisms of behaviour have tended to call themselves neuroethologists or neurobiologists. Those interested in adaptation have called themselves behavioural ecologists. Each group has developed its own terminology and founded its own journals and each has concentrated on a sub-set of what used to be called ethology. What has come to be known as 'sociobiology' – the adaptive study of social behaviour – actually refers to a sub-set of behavioural ecology which is in turn a sub-set of ethology, and one that Tinbergen not only knew about but largely pioneered, except that he did not use that name.

All subjects should develop, of course, and a few changes of name would, by themselves, be no bad thing (and would indeed be a welcome reflection of a growing interest in animal behaviour) were it not for the fact that they are indicative of a change in attitude that has in practice turned out to have disadvantages. Behavioural ecologists and sociobiologists, by concentrating on adaptation, have neglected the study of mechanism to such an extent that theories of sexual selection, for example, are discussed without reference to the basic way in which a receiver's sense organs work. Discoveries made years ago by the classical ethologists on the response of animals to 'sign stimuli' and 'supernormal' stimuli have been neglected and then only recently rediscovered as if they were something brand new. Discussions of the function of 'conspicuous' plumage in birds have gone on without taking into account that what is conspicuous or inconspicuous to a human may not be so to another bird.

The emphasis on adaptation to the exclusion of mechanism has had other unfortunate consequences, too. Many of the practical problems that arise in connection with animal behaviour – such as how to keep birds off airfields or how to reintroduce captive-bred animals into the wild – are basically questions about mechanism and development of behaviour, not primarily about adaptation and fitness. The answers that people want in their day-to-day interactions with animals in zoos, farms, wildlife reserves or in their own homes are only distantly related to optimality and gene strategy and much more likely to be found by studying how animals respond to various stimuli in their environment and whether their behaviour is innate or learnt. Because such questions have been relatively neglected, the whole subject of animal behaviour appears much more rarefied and far removed from practical issues than might otherwise have been the case. If it is to appear useful, the study of animal behaviour will have to return to its four-question roots, quite apart from the fact, intellectually, it would benefit from asking 'How does it work?' as well as 'What is it for?'

There are probably two reasons why there has been this relative neglect of the machinery of behaviour. The first is, as we have seen, the spectacular success of evolutionary explanations of behaviour in the light of the 'selfish gene' reinterpretation of Darwinism that took place in the 1970s. W. D. Hamilton's papers on inclusive fitness and R. Trivers's (1971) paper on reciprocal altruism meant that a whole host of apparently altruistic behaviour could be seen as the way in which genes, often by the most surprising routes, could get themselves into the next generation. There is something very attractive about a unifying theory that can be used to explain a diverse body of evidence – and that was precisely what the study of the mechanism of behaviour lacked.

In fact, it was worse than that. The second reason for the neglect of the study of mechanism is that ethology had once had a grand theory of how the machinery of behaviour worked and then, sometime in the 1960s, it lost it and never found a replacement (Bateson, 1993). The Lorenz–Tinbergen theory of instinct was such a theory and when it was toppled, there was a vacuum which even now has not been filled. To understand why the mechanisms of behaviour have been until relatively recently neglected, we need to take a brief look at the rise and fall of the grand theory of instinct. Then we will look at some of the different approaches, such as circuit diagrams, black boxes and neural networks, that have more recently been used with greater or lesser success to study the mechanisms of behaviour.

What is 'instinct' and what is wrong with it?

Classical ethology, under the influence of Konrad Lorenz and Niko Tinbergen, saw animal behaviour as 'driven from within' – the literal meaning of instinct. Lorenz's (1950) view was that the nervous system spontaneously produced energy that then drove the animal to perform particular behaviour patterns. Tinbergen (1951), more tentatively, wrote of the 'internal factors which determine the "motivation" of an animal, the activation of its instincts' (p. 57).

At the time, the idea was quite revolutionary. The prevailing opinion in physiology was that animal behaviour should be seen as a series of reflexes, with the animal reacting to a series of external stimuli. Coordinated behaviour was seen to be the result of a chain of these reflexes in which performance of the first small step in the movement would stimulate the sense organs to activate muscles to produce the next step and so on. The animal was thus seen to be essentially *re*acting to stimuli in its environment, rather than being spontaneously active of its own accord.

Here were Lorenz and Tinbergen putting the emphasis quite the other way around. They saw the animal's nervous system as the active party, energizing and driving the behaviour. They did also acknowledge some role for stimuli outside the animal, because it is quite obvious that animals are affected by things around them, but for Lorenz in particular, it was the build up of instinctive energy inside the animal that at one and the same time drove the behaviour and also lowered the threshold of response to external stimuli.

Lorenz believed that the longer the time an animal went without doing a particular kind of behaviour, the higher the levels of this nervous energy would become and the more likely the animal would be to perform the behaviour. Animals would become more and more likely to feed or be aggressive, for example, the longer they had been without the opportunity to perform these behaviours until eventually they might start doing them without any of the usual external stimuli being present. They might start fighting or mating with their food dish, for example, or even trying to bury nuts in imaginary earth. Exactly how many instincts there were was left open but they corresponded to broad categories of behaviour such as eating, fighting, parental behaviour and sleep (Tinbergen, 1951) rather than very specific ones such as different kinds of fighting.

The appeal of the Lorenz–Tinbergen view of instinct was that it attempted to give a single explanation for very different kinds of behaviour – in broad outline at least. Obviously the causal mechanisms for sleep and fighting would be different in detail. There would be different hormones released, different nervous mechanisms at work and so on, but all these diverse mechanisms were seen as having important things in common. They were all held to share the common characteristics of being driven from within the animal, with energy being built up and released when the animal encountered the right external stimuli. Instincts, generated from within, underlay everything.

Unfortunately, all notions of instinct suffer from the same fatal flaw. It is not just Lorenz's idea of instinct as 'energy' or his celebrated analogy of the nervous system with cisterns accumulating water that are at fault, but any concept of an energizing force inside the animal, whether it is called 'instinct', 'drive', 'action specific energy' or what. All of them suggest pent-up physical energy, like the exploding gases in the cylinders of an internal combustion engine.

But instinctive energy does not power muscles. Chemical energy does that. So 'instinctive energy', if it means anything, must refer to something else, perhaps the sum total of internal causal factors (hormones, neural responsiveness, etc.) that exist at any one time. We could talk about an animal having a high level or a low level of these factors, but even a high level would not mean that the animal was necessarily going to be energetic or rush round in a frenzy. A high level of causal factors for sleep, for example, results in inactivity. A sleep 'drive' 'energizing' the animal becomes a contradiction in terms and shows up the error of careless thinking about energy concepts. Animals do not have *energy* accumulating inside them. Various changes may take place in their bodies and these changes may (or may not, as the case may be) get progressively larger as time elapses. But they are not like the forces of gases in a cylinder forcing the piston to move. To think of animals like this is to make a serious error about the nature of the very complex processes that are going on inside their bodies. If we want to understand these processes, it is better not to be misled into thinking that there are instincts 'propelling' behaviour, attractive and superficially plausible though that idea may at first seem.

Hinde (1960) led the attack on 'drive' and 'energy'. 'Instinct' suffered from all the ills that he revealed in 'drive' plus another of its own: its implications of innateness. We have already dealt, in Chapter 4, with the controversies that have surrounded that idea. Instinctive behaviour, with its 'inborn' as well as its 'driven from within' elements under attack, should have been dead and buried.

But it has not disappeared entirely, however much its critics would like to see it go. It has such a powerful intuitive appeal that many people outside ethology still refer to instinctive behaviour and are under the impression that this is what the study of animal behaviour is all about. More interestingly, some other concepts that are still to be found in the scientific literature owe their continued use to the assumption that there are instincts that have at least some of the properties attributed to them by Tinbergen and Lorenz. These include 'displacement activities' (the idea that energy can be 'displaced' from one behaviour system to another and so give rise to inappropriate behaviour such as a man in a quandary scratching his head or a gull breaking off a fight to pull up grass) and 'vacuum activities' (the idea animals will perform behaviour 'in a vacuum' or with none of the right external stimuli if their motivation to do them becomes high enough).

In general, however, the concept of instinctive energy inside an animal driving it to behave in certain ways is now no longer accepted. People have had to look elsewhere for their explanations of how animals behave but have failed to find a theory big enough to take the place of the Lorenz–Tinbergen theory of instinct. They have instead become embroiled in a somewhat sterile controversy about the right way to go about studying behaviour and, in so doing, have once again fuelled controversy and given rise to confusion.

The controversy is now not so much about the value of a particular model (the idea of one model being applicable to all behaviour being more or less abandoned as an impossible goal) as of what might be the best strategy or approach for studying the immensely complex control of behaviour. Animals – even ones with relatively small nervous systems – are very complicated, both in terms of the diversity of behaviour they produce and the number of parts that make up the machinery controlling behaviour.

One view – that of neurobiologists – is that the best way to understand what is going on is to look inside the animal's body and to try to see how the various cells and organs function. The other view – the 'whole animal' or 'black box' view – is that knowing how individual parts work is not and never will be sufficient to understand behaviour and that what is needed is to study the behaviour of whole, intact animals without opening them up.

At times, this controversy has become quite acrimonious, with each side claiming that their approach is not only the best but actually the only sensible way to study animal behaviour. To see our way through this controversy, we will now look at what these various approaches have – and have not – told us.

THE DIRECT APPROACH: MAPPING THE NERVOUS SYSTEM

A major breakthrough in the understanding of the way the nervous system gives rise to behaviour came with the discovery that single nerve cells, at least in invertebrate animals, could be identified and shown to have certain characteristic roles in the production of behaviour. This idea, first put forward by Wiersma in the early 1950s, did not receive general acceptance until fifteen or so years later when the use of distinctive coloured dye enabled the connections of a given nerve cell to be traced clearly for the first time. Single neurons, far from being anonymous entities in a crowd of other neurons, began to be seen as having an identity of their own. Not only was a nerve cell revealed as having certain specific connections to other cells, but a very similar cell, with similar connections, would be found in other individuals of the same species.

It was not surprising to find that all individual insects of a particular species have six legs and similarly shaped jaws, but it was something of a surprise to find that they all had many individual cells of their nervous systems the same too. The nerve cells and their connections could be mapped like the circuit diagrams of a computer and the maps then applied to all individuals of a species. For example, the circuit diagram controlling the way a locust jumps away from a visual stimulus was drawn up in considerable detail, from the cell at the base of each eye that picks up any movement in the locust's visual field, through several identified interneurons right through to the motorneurons responsible for activating the muscles that extend the legs for a jump (Burrows & Rowell, 1973; O'Shea & Williams, 1974).

By the 1990s, however, neurobiologists were sounding distinctly less confident. 'It is slowly becoming appreciated', wrote Altman & Kein (1990) 'that the view of the nervous system down the barrel of a microelectrode is rather limited.' The reason for these creeping doubts was that although 'circuits', usually of between 20 and 30 neurons, had been described for a number of different invertebrates, these did not seem to give a complete picture of the control of the behaviour that resulted. The neurons in the circuits were found to be connected not just to each other but to hundreds of other neurons as well and many neurons were involved in more than one circuit. When the sea-slug *Aplysia* withdraws its gills in response to being touched, over 300 neurons in its abdominal ganglion are active, far more than had been expected. When a leech bends its body, also in response to being touched, a whole network of interneurons comes into play (Lockery *et al.*, 1989). The leech has sensory neurons that respond to touch and motor neurons that control particular muscles, but most of the interneurons respond to more than one kind of sensory input and most of them have effects on many different motorneurons. The behaviour of the leech seems to be determined not so much by specific circuits as by a network of interneurons, active and inactive in different ways at different times (Lockery and Sejnowski, 1993). Even for animals as simple as a leech, neurobiologists are having to revise their views of the way the nervous system works and think of nerve cells not as localized entities but as part of a distributed processing and control system that may have extensive connections outside its obvious sphere of influence. When it comes to vertebrates, the rethinking has been even more drastic. John *et al.* (1986) reported that when a human being is performing a simple visual learning task, an astonishing 5–100 million neurons are active throughout the brain. Simple circuits are clearly out of the question. It is the connections between neurons that hold the key.

WHOLE ANIMALS, 'BLACK BOXES' AND MOTIVATION

For some people, the complexity of the vertebrate nervous system is so great that a completely different approach to understanding mechanisms of behaviour is needed. For them, the way forward is not to look inside animal bodies as a neurobiologist would but to treat them as a whole and to try to work out what their internal mechanisms are from the way they behave. When it became apparent that the particular model proposed by Lorenz and Tinbergen failed to explain a great deal of animal behaviour, they cast around for other models and other approaches. One of these was to employ terms like 'motivation' and to treat an animal as a 'black box', an initially mysterious object that is not to be opened but whose workings can be deduced from what it is capable of doing. There are two different reasons for adopting this 'whole animal' approach.

The first is that it is a kind of 'poor man's physiology'. We are such a long way away from being able to give a complete neurophysiological account of even the simplest behaviour of the simplest animal that it is necessary to employ terms like 'motivation' to cover the fact that we are ignorant of much of the physiological basis of, say, the bee dance, fighting in red deer or nest-building in birds. All that is meant by motivation in this context is the as yet unknown internal causal factors which lead animals to show these behaviours. We would like to know what these causal factors are – in other words we would like to be physiologists if we could – but until neurobiology catches up with what ethologists want to know about and makes the term 'motivation' completely redundant, we will continue to use it.

The other approach is more aggressive. Instead of seeing the 'whole animal' approach as an unsatisfactory substitute for the real thing (i.e. neurobiology), it is seen as the only way to study certain phenomena. Unless you understand what an animal or machine can do, poking around inside it is going to be fairly meaningless. Even a complete circuit diagram of a computer would not tell us that it could play chess or recognize shapes because understanding how a computer plays chess involves understanding the program, the software that could in principle be run on many different computers with different details in their wiring. As Oatley (1978) put it, we would not go about finding how a computer plays chess by 'cutting cable tracts, removing circuit boards and for larger scale operations removing substantial quantities of the computer with a shovel' (p. 25) which he claims is the equivalent of what neurobiologists do to the nervous system.

Certainly, we would not find out a great deal about how an animal's behaviour is controlled if the first thing we did was to record from its brain with no knowledge at all of what it could do – that it could outmanœuvre a hawk in flight, for example, or tell the difference between the distortions in an electric field caused by objects made of different materials as electric fish can. By treating animals as black boxes and describing in detail what they do, it is often possible to arrive at 'software' explanations of behaviour that are completely valid in their own right.

Genetics provides an interesting parallel here. Although there have of course been major advances in molecular genetics that could only have come about through probing the genetic material itself, there are many other discoveries that were made, and perhaps could only have been made, through a black box approach. It was not molecular genetics that told us about segregation and sex-linkage, about crossing over and dominance. It was 'black-box genetics'. Mendel deduced the existence of factors inside the germ cells of organisms with no knowledge at all of what their material basis might be. He even inferred that the factors segregated independently and did not contaminate one another. Morgan (1911) and then Sturtevant (1913) took black box genetics even further and used it to map the position of genes along a chromosome.

It is with hope of some similar success that ethologists have taken a black box approach to the study of animal behaviour. Unfortunately their path has not been without its pitfalls and confusions, particularly when they have used the concept of motivation.

What is 'motivation' and where does it get us?

The idea of motivation is used to explain the fact that animals do not always respond in the same way to the same external stimulus. If we watch a bird, we might see that at one time it is feeding and then some time later it stops and goes to sleep. There is still plenty of food available and nothing in the environment appears to have changed so it must be something inside the animal that has changed and caused the shift from feeding to sleeping. And it is not just one specific kind of behaviour that is affected. The animal has stopped pecking at food and it has also stopped turning over leaves and actively walking round with its head close to the ground. It has now switched to a completely different set of actions – closing its eyes, fluffing its feathers and so on.

For the moment we may be quite ignorant about the physiological basis of the changes going on inside the animal that cause it to switch from one set of activities to another but we can nevertheless infer something about what those changes may be like. They must, for instance, be reversible because we observe that after a period of sleep the animal goes back to its previous behaviour of feeding. They must also be changes that affect not just one behaviour pattern but a whole range of actions concerned with obtaining food or resting. We might even be able to infer something about what they might be from the fact that the time spent sleeping depends on how much the animal had previously eaten, what it was eating, the time of day and the fact that the feeding and sleeping seem to be done according to some sort of rhythm.

These changes – whatever they are – are referred to as changes in the animal's 'motivation'. Motivational changes are usually distinguished from those associated with injury or fatigue and also from longer-term and more permanent changes due to learning or maturation. The bird in our example was undergoing motivational changes – short-term reversible alterations in its internal state. Now, as we have already seen,'motivation' is to a certain extent a declaration of ignorance. If we knew exactly how the behaviour was produced, we would have no need to use the term at all. We would all be neurobiologists. So the question we have to ask is, what use is served by cloaking our ignorance in such a deliberately vague term and pretending to explain something which we have not really explained at all?

The first use is when it leads directly to the discovery of physiological mechanisms that would not otherwise have even been looked for. For example, Simpson & Ludlow (1986) constructed a motivational model of feeding behaviour in locusts in which the probability of feeding was given by an interaction of rhythmically fluctuating feeding motivation with a threshold that varied both with the time since the last meal and with how big it had been. Once they had shown that their motivational model successfully described the feeding behaviour of real locusts, they then went on to investigate the physiological basis of the various parts of the model and found, for instance, that defecation, probably mediated through a change in the firing rate of the stretch receptors in the hind gut, was one of the major 'motivational' factors determining when a locust started to feed.

A second use for motivational analyses is when they show that the control of behaviour is more complex than appears at first sight and so alerts physiologists to the scale of what they have to explain. For example, rats drink different amounts of water at different times and so it would be easy to assume that this was due to fluctuations in a single 'drinking motivation'. However, Miller (1957) showed by a purely 'whole animal' approach that this could not be true. When rats were injected with salt (which might be expected to increase their drinking motivation), the amount of water they drank went up for 3 hours and then levelled off, suggesting they were no more 'motivated' to drink after 6 hours than after 3 hours. But the amount of quinine they would tolerate in the drinking water was much greater after 6 hours than after 3 hours, suggesting that by this measure, they were much more 'motivated' to drink after 6 hours. Drinking is not, therefore, caused by simple changes in one variable but must involve many different pathways.

Tests of Optimal Foraging Theory (Chapter 2) can also be seen as investigations of behaviour mechanisms carried out on whole animals even though this may not have been the declared reason for doing them. It is quite clear from the huge number of such studies that have been carried out that animals do not feed just because they have a nutritional deficit and suitable food is in front of them. Rather, their feeding behaviour is influenced by such things as how far they would have to travel to the next clump of food, the amount of energy they would gain from one food item as opposed to another, the nutrient content and taste of the food, whether they have hungry young to feed and whether the food source is reliable or risky (Caraco *et al.*, 1990). Elgar (1986a) showed that whether house sparrows start feeding was even dependent on whether the food was in one lump or the same amount divided into smaller pieces that could be shared with other sparrows.

This value of whole animal studies in warning neurobiologists of the complexity of the mechanisms underlying behaviour is even more apparent when we look at interactions between motivations for different behaviours, for example at the way in which feeding motivation interacts with that for drinking, mating or escaping from predators. (There is a terminological split here. Some people make a distinction between internally generated 'motivation' for feeding and drinking on the one hand and externally stimulated 'emotional states' such as those produced by the appearance of predators on the other; others argue that the control of behaviour such as fear, aggression and sexual behaviour is not fundamentally different from that of feeding and drinking (Toates, 1986) and that they can equally be described as being 'motivated'. I shall follow the latter view.)

Elgar (1986a, b) showed that, in addition to the divisibility of the food, whether or not a sparrow feeds is due to a delicate balance between its motivation for feeding and its fear of predators. With a high motivation for escaping (when a human being was close to the food source) feeding only took place when there were other sparrows present since other birds, with many pairs of eyes, provide a degree of mutual protection and advance warning against predators. If no other sparrows were present, the bird that had discovered the food would not feed itself initially despite being motivated to feed, but would call repeatedly until joined by a sufficient number of companions. Motivation for feeding, escaping and being social were therefore intimately linked together and any physiological account of feeding behaviour that did not take this into account would simply be unable to 'explain' the observed feeding behaviour of wild animals.

It is also possible to use a whole animal approach to investigate how different motivations for different behaviour interact. When they are incubating eggs, female junglefowl are motivated both to feed and to stay on the nest. Normally, junglefowl hens spend over 60 per cent of the daylight hours scratching in leaf litter and feeding but when they have eggs, they sit almost continuously and generally leave the nest only once or twice a day for 5–10 minutes to feed and drink. Over the 21 days of incubation, they may lose up to 17 per cent of their body weight but, curiously, even if they are provided with food at the nest so that they could eat without leaving the eggs, they still refuse to eat and still lose weight (Sherry *et al.*, 1980). This suggests that the motivation for incubation is suppressing that for feeding during this period. However, feeding motivation is still affecting the behaviour to some extent because if hens are deprived of food during one of their 'meal breaks', they return to the nest but leave again earlier and eat more when food does become available than undeprived hens (Hogan, 1989). The motivation for incubation is also affected by the time of day as the later in the day that food becomes available, the less likely hens are to leave the nest at all to get it. If food is only available in the late afternoon, hens will go without food altogether for 48–72 hours rather than leave the nest. Eventually, however, they will leave to feed. Motivation for incubation and that for feeding therefore seem to be in some sense in competition, with that for incubation dominating most but not all of the time. We do not yet know quite what form this competition takes inside the animal or quite how one motivation can dominate another but the whole animal approach has indicated the kinds of interaction that must be going on. It has paved the way for a more detailed physiological study later on.

NEURAL NETS AND THE CONNECTIONIST REVOLUTION

There are limits to a behavioural approach to investigating the machinery of behaviour. There comes a time in the analysis of any system when it becomes necessary to discover the physical embodiment of the entities or processes that have been deduced. Genetics has moved on from its 'black box' days and ethology must too. It has in fact become more and more apparent that treating animals just as 'black boxes' and refusing to take any interest in what is going on inside can be just as misleading as thinking that circuit diagrams are the answer to everything – and for an interestingly similar reason. Circuit diagrams were seen to have their limitations when it became apparent that nervous systems – even those of 'simple' animals – were probably organized not as circuits but as whole networks of interconnected neurons. Black boxes began to seem distinctly old-fashioned when psychologists (who were responsible for the 'black-box' terminology in the first place) realized that human information processing should not, as they had previously thought, be seen as taking place inside a series of clearly defined boxes each with a separate function. Brains did not work like that. They did not go through a series of instructions like a linear computer program ('If *x* then *y* . . .'). Rather, they operated as a network of interconnected cells with information being processed in parallel and in all parts of the network at once, like a different sort of computer program, with what computer scientists call a content-addressable memory (Quinlan, 1991).

There is not space in this book to explain the 'connectionist' revolution that has overtaken psychology in the last 10 years or so and which is gradually extending to other areas of behaviour study, including, as we have seen, the behaviour of leeches. For the purposes of this chapter and the implications for the debate about the best way to study the machinery of behaviour, the conclusion can be stated quite simply. Brains matter. If brains – human or non-human – function like conventional serial processing computers then they will give rise to one sort of behaviour. If, however, they function like parallel distributed processors then they will do different things and give rise to different sorts of behaviour. Ethologists can no longer afford to ignore what the nervous system is actually doing or make lofty statements about only being interested in 'software' and 'whole animal' explanations because – to pursue the computer analogy further – it does matter what sort of machine is running the program. On the other hand, neurobiologists have to face up to the complexity of the neural connections that underlie what can at first sight look like a 'simple' piece of behaviour and think of networks rather than circuits. Perhaps neural networks will bring workers in different fields together and see a welcome end to the chauvinism (both neurobiological and ethological) that has tended to get in the way of linking physiological and behavioural studies and unnecessarily fuelled the debate about how 'best' to study the mechanisms of behaviour. (There is not just one best way. We need to know what the behaviour is and what has to be explained just as much as we need explanations for what causes it.)

Conclusions

The study of animal behaviour has, over recent years, become fragmented into behavioural ecology, where the emphasis is on adaptation and evolution and neurobiology, where mechanisms of behaviour are studied from a physiological point of view. The importance of integrating both approaches is only gradually being realized or rather, rediscovered. Classical ethology, under the influence of Tinbergen and Lorenz, was built on the idea that questions about adaptation and questions about mechanism should go hand in hand. Even though the classical ideas about instinct are no longer considered to be useful in understanding how behaviour works, the earlier emphasis on taking a broad view of what it means to 'explain' behaviour is beginning to come back into fashion. As we saw in the last chapter, for example, studies on the evolution of sexual preferences are being illuminated by studies of the mechanisms of mate choice. Behavioural ecologists and neurobiologists are now taking more notice of each other's findings and ideas, although there is still a long way to go.

One of the reasons why the links between neurobiology and ethology are weaker than they might be is that only rarely are the same animals studied and comparable questions asked. The most informative studies are those, such as prey-catching in toads (Ewert, 1987), hearing in barn owls (Konishi, 1986), song-learning in birds (Konishi, 1985) and echo location in bats, where an intensive study of behaviour is followed up with neurophysiological studies on the same animal. In the case of bats, for example, behavioural studies were used to show that bats use the delay between the sound they make themselves and the echo that comes back to them from surrounding objects to determine how far away things are (Simmons & Vernon, 1971). Neurophysiological recordings from the auditory cortex of bats then revealed a class of cells that were 'echo detectors' and only responded to echoes, not to the much louder sound of the bats' initial call (Feng *et al.*, 1978). They would not respond even to an echo if there had not just been a loud call immediately beforehand. Some of the cells responded most vigorously if there was a delay of 4.5 ms between call and echo, others if it were 20 ms. The difference between 4.5 and 20 ms corresponds to a difference in target distance of 0.7 and 3.5 m so how far away an object is from the bat is thus, at least partly, encoded in different places in its brain.

All too often, however, neurobiologists study animals such as leeches or sea slugs, that have behaviour which ethologists think, to be quite honest, is boring, whereas ethologists study birds or mammals where the neurobiology is poorly understood. In many ways, the hope of linking the two fields has been receding rather than increasing lately because as they work away at their separate animals, neurobiologists find that even their 'simple' nervous systems turn out not to be that simple in the way they operate and ethologists discover that the behaviour of animals is far more complex and influenced by more factors than they had imagined. For all that we have learned about the way in which the nervous system works, however, we should not lose sight of the fact that animals are far more complicated and versatile than any machine so far made by man. We are – both neurobiologists and ethologists alike – a very long way away from a complete explanation of even one example of animal behaviour. And in trying to gain such an explanation, we are in pursuit of one of the most extraordinary and remarkable phenomena seen anywhere on earth.

Further Reading

Young (1989) gives an excellent account of the links between behaviour and neurobiology and Colgan (1989) takes a motivational approach. Hinton (1992) shows how learning occurs in neural networks and a whole issue of *American Zoologist* (**33** (1), 1993) is devoted to the application of neural network thinking to animal behaviour.

8

COGNITIVE ETHOLOGY

Homing pigeons (*Columba livia*) at release. Photograph by Tim Guilford.

Of all the words in current use to describe the complexities of animal behaviour, the one that promises the most and in the end delivers the least is probably 'cognitive'. It is one of those words, like 'innate', that different people use in different ways to mean different things. Sometimes it is taken to mean 'mental processes' (McFarland, 1989), often with the implication that those mental processes are conscious (Griffin, 1993; Bekoff, 1993). At other times, it means something more down to earth, with no necessary implication of conscious awareness. McClean & Rhodes (1991), for example, use the term 'cognitive' to describe a huge array of animal behaviour, excluding only very simple, fixed responses. Thus, the way in which many animals respond to their predators is, for them, an example of cognitive behaviour because stored knowledge of the world (either innate or acquired) is used in a flexible way to interpret what is seen and to choose an appropriate response. Depending on circumstances, the same animal might show mobbing, alarm calling, hiding or distraction behaviour, indicating that it had the capacity to assess the situation and adjust its behaviour accordingly. Only an animal that had a very simple enemy-recognition system (to a looming shadow, say) and one possible response (withdraw into burrow) would not, in their terms, be showing cognitive behaviour. It is the flexibility of the response, not whether it is carried out consciously or unconsciously, that earns it the title 'cognitive'.

A similarly broad definition is given by Tooby & Cosmides (1992). They argue that it is possible to describe all but the very simplest reflexes of animals as 'cognitive' because cognitive refers not to a level of complexity in the animal but to the approach of the person studying it. 'Reasoning, emotion, motivation, and motor control can all be described in cognitive terms, whether the processes that give rise to them are conscious or unconscious, simple or complex' (p. 65). To them, as to McClean and Rhodes, what matters is how information is received, stored and used, not whether these processes are carried out consciously. A 'cognitive' approach is one that is concerned with how these processes take place. Even a herring gull chick pecking at the red spot on its parent's bill and stimulating it to regurgitate food would be described in cognitive terms because questions can be asked about how the chick recognizes the parent and how it uses information from the environment to direct its pecking behaviour.

Using 'cognitive' in this sense of 'information-processing mechanisms' would, of course specifically state that computers are cognitive because they are, *par excellence*, machines that input, process and store information. And putting it this way makes it clear that the most useful sense of 'cognitive' (the one in current use in, for example, cognitive psychology or cognitive neuroscience) is to make a major distinction between 'information processing' (done unequivocally by humans, computers and many non-human animals) and 'consciousness' (possessed but not always used by humans and attributed much more controversially to anything else). Cognitive psychology is concerned with information processing and how human brains represent the world. Logically, cognitive ethology ought to be doing the same thing for non-human brains but it has unfortunately got itself tied up with the issues of consciousness to such an extent that for many people, cognitive ethology and conscious ethology are virtually synonymous.

It is important to realize, however, that 'cognitive' and 'conscious' are not the same thing. 'Cognitive', as we have just seen, refers to the processes by which information is perceived, stored and processed. It therefore covers a range of processes such as pattern recognition, attention and memory, all of which can be carried out by machines as well as people and other animals. 'Conscious', on the other hand, refers to the subjective awareness that, for us, often accompanies this information processing (as when we experience a conscious recognition of a long-lost friend). This conscious awareness of what we see and do may or may not be present in other species and we will return to the issue of whether it is in the next chapter. But for the present chapter 'cognitive' will mean cognitive and not conscious and we will be looking at some of the controversies surrounding the question 'How do animal brains represent the world?' As we will see, this question is difficult enough in itself without it being contaminated with the even more difficult issue of consciousness.

What is a representation?

There is nothing very mysterious about the idea that some aspects of the world are represented in the brain of an animal, or in a machine for that matter. As Gallistel (1990) put it, all it means is that 'there is a functioning isomorphism between some aspect of the environment and a brain process that adapts the animal's behaviour to it'. There must be something going on in the brain of the animal, in other words, that corresponds to something going on in the world outside in such a way that the animal is able to behave appropriately. For example, if an animal has a representation of one of its major predators, that would mean that it could not only recognize the predator when it saw it, but could adapt its own behaviour to give the appropriate reaction depending on whether the predator was approaching or moving away, had seen it or appeared not to have noticed it and so on.

The advantage of having such a representation of the world, is the flexibility of response that it gives to the machine or animal possessing it. A non-cognitive animal without an internal representation might achieve a certain degree of flexibility by having a long list of rules such as 'If predator stimulus appears, do action A; if action A is not followed within 15 s by removal of stimulus, do action B and so on.' A cognitive animal, one that had an internal representation of the predator, on the other hand, could achieve a far higher degree of flexibility. For example, it might be able to weigh up what the predator was likely to do and adjust its own behaviour accordingly. It might show mobbing, escape, hiding or freezing to ostensibly the same stimulus. It might even show behaviour it had never shown before (McClean & Rhodes, 1991).

An even better example of how an internal representation aids flexibility of response is given by Cheney & Seyfarth (1988) in one of their many illuminating studies of vervet monkeys in the wild. Vervet monkeys have two different calls that signifiy the presence of a strange group of other vervets – a 'wrr' call and a 'chutter' call. Both calls have the effect of alerting other monkeys in the group that there are strangers about and making them look around but they are used in slightly different circumstances. What Cheney and Seyfarth did was to make a tape-recording of the 'wrr' call of one particular individual monkey and then play it over and over again to the members of its troop so that they became habituated to it. After hearing it eight or nine times, they no longer looked around for other monkeys because they realised that the call signified nothing. The experimenters had effectively made the individual whose call was being repeatedly played appear to be 'crying wolf'.

Cheney and Seyfarth then played the 'chutter' call of the same individual and found that for this call, too, the other monkeys took no notice – the habituation to one of its calls had been generalized to the other. However – and this was the crucial part of the experiment – when the 'wrr' or 'chutter' calls of *other* individuals were played, the group of monkeys responded strongly, suggesting that what they had learnt was that the calls of one particular monkey were unreliable, not that 'wrr' and 'chutter' calls in general should not be responded to. This means that the monkeys could not only distinguish between calls of different individuals but also generalize from two different sounding calls from the same individual. This is good evidence that they had an internal representation of a particular monkey as an unreliable signaller of the presence of strangers that enabled them to adjust their behaviour immediately. They did not have to learn that the chutter call of that particular monkey was unreliable. They inferred it from what they had already learnt about its 'wrr' call. They seemed, then, to have a representation of that particular monkey as 'unreliable', not just a learnt response of habituation to a particular call.

This is, of course, an example of a sophisticated internal representation, one that we would not necessarily expect to be found in all animals. But there is one kind of representation that seems to be almost universal among fish, birds, mammals and even insects – namely, the internal representation of space.

Cognitive maps

To say that an animal has a 'cognitive map' might be taken to imply that the animal was doing something extremely clever in its head, for example, that it had a complete internal representation of its world and could manœuvre itself around by working out internally the best route to go. Such, indeed, was the meaning attributed by Tolman (1932) to the idea that an animal might have a 'map in the head'. The test of this would be to train an animal, a rat, say, to run through a maze to find food and then block off the rat's usual route. If the rat could immediately (i.e. without further exploration) choose another correct route then the conclusion would be that it must have worked out what to do 'in its head' from its previous knowledge of the maze. It could only do this if it had a considerable knowledge of the various parts of the maze and their spatial relationship to each other.

However, although such a rat, if it performed such a task, would undoubtedly be said to have a 'cognitive map', more recent usages of the term would attribute maps to many other animals with rather less complex behaviour. Gallistel (1990), for instance, extends the definition so broadly that he claims that most animals that move through space have cognitive maps. Although this sounds a bit sweeping, it is in line with the definition of 'cognitive' as 'representational' that we have adopted in this chapter. As long as the spatial positions of objects are in some way represented in the brain of an animal, that animal can be said to have a cognitive map. The interesting question then becomes what sort of map it has.

Some cognitive maps are 'weak' in the sense that only some geometric relationships are represented. The digger wasp females studied by Tinbergen (1932) have such weak maps. Tinbergen found that if he moved a ring of pine-cones immediately surrounding a wasp's nest, she was unable to locate the nest correctly. She went instead to the centre of the ring of pine-cones where the nest no longer was. The wasp had thus noticed something about the spatial relationship between certain landmarks and the nest, but could easily be deceived by moving a small number of landmarks a short distance away, showing that other apparently obvious features in the landscape were not noticed at all.

Other animals, such as the rat working out which alternative route to take when its usual one is blocked have what Gallistel calls 'strong' cognitive maps, that is, ones which contain much information about spatial relationships between objects. In fact, the only animals that Gallistel would say do not have cognitive maps at all are those that move from one place to another using beacons. An animal that can sense its goal from its starting point and simply heads towards this single beacon stimulus can do so with no information whatsoever about spatial relationships between landmarks. It responds directly to this stimulus, whereas any animal that is sensitive to spatial relationships must obviously be responding to at least two objects or points.

For Gallistel, then, the interesting question is not 'Do animals have cognitive maps' (because as far as he is concerned cognitive maps are nearly ubiquitous among animals) but rather, 'What sort of information does the animal have in its map?' More specifically, it is important to ask 'What sort of geometric relationships does the map encode?' The information might be sufficient to enable an animal to find its way home even when placed in a completely novel part of its environment or when forced, like Tolman's rat, to work out a new route to a goal. On the other hand, like the digger wasp, it might be much more limited and mean that the animal would be unable to find its way home even with relatively minor disruptions.

At this point, we encounter a potentially confusing difference of opinion over terminology. We have taken the mystery out of the term 'cognitive map' by making it clear that it means nothing more than 'representation of spatial relationships'. This down-to-earth definition also implies that 'cognitive map' explains little in itself. The interesting part comes next when we ask what sort of cognitive map an animal has and what sort of spatial abilities it possesses. Unfortunately, the classification of the different sorts of cognitive maps animals might have varies with different authors. Particular confusion has centred on what is called 'pilotage' and what is called 'navigation'.

For obvious reasons connected with the spectacular nature of their migratory journeys and feats of homing, much of the work on animal maps and spatial representations has been done on birds, particularly homing pigeons. Classically (e.g. Baker, 1984), a distinction was made between 'orientation' (the ability to maintain a compass direction), 'pilotage' (the ability to find one's way to a known goal over familiar terrain using known landmarks and true 'navigation' (the ability to find one's way to a known goal over unfamiliar terrain, i.e. where there were no known landmarks). In practice, the distinction between pilotage and navigation could be made by taking birds such as pigeons to completely unfamiliar areas and seeing whether they could still find their way home even when no familiar landmarks were present. Similarly, the distinction between orientation and navigation could be made using the so-called 'displacement' experi-

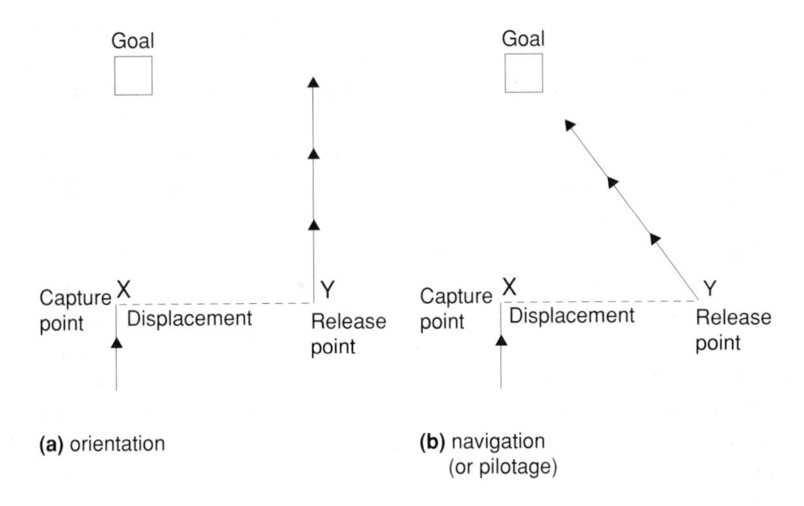

Fig. 3 The difference between orientation and navigation. An animal travelling towards a goal such as a home loft is captured at point X, displaced along a line perpendicular to its line of travel and then released at point Y. If, when released, it continues in the same direction as before, it is showing compass orientation. If, however, it changes direction and heads towards the goal, it is showing either pilotage (if relying on familiar landmarks) or navigation (if the terrain is unfamiliar).

ment. Birds would be captured *en route* and then transported in a direction perpendicular to that in which they had been moving (Fig. 3). If, when released, they continued in the same direction as before and ended up in the wrong place, they would be said to be showing orientation but if after release they changed direction so as to compensate for the displacement and reached the goal, they would be said to be showing true navigation (or pilotage if the area were familiar).

Such distinctions have been made with real animals. Perdeck (1958) did a classic experiment in which he captured both young and adult starlings on their migratory route through the Netherlands and then took them to Switzerland. When he released them, the young birds continued to migrate in the same south-westerly direction they had been following before capture and ended up in Spain instead of in Britain and northern France, which are their normal wintering grounds. The adults, on the other hand, changed direction and ended up in the right place despite the disruption of their capture and displacement. In other words, the young inexperienced birds were showing compass orientation, whereas the adults were using a different mechanism (true navigation or just possibly pilotage).

More recently, however, Papi (1992) has drawn up a classification of animal movements that cuts across many of the earlier ones and, unless care is taken, could lead to considerable confusion because of its differences with some earlier terminology. To begin with, Papi's definition of 'navigation' is considerably broader than that previously offered by Baker (1984). Papi includes under 'navigation' not just the ability to find one's way over unfamiliar terrain (true navigation in Baker's sense) but also cases where birds use specific familiar local cues, thus including elements of what Baker would call 'pilotage'. A bird using a 'map' made up of a set of familiar visual, olfactory or sound cues would be navigating in Papi's sense but showing pilotage in Baker's. On the other hand, Papi excludes from the definition of pilotage any case where an animal uses landmarks in conjunction with a compass. This is potentially even more confusing because Gallistel (1990) specifically argues that pilotage involves an interaction between compass orientation and landmarks. So what Gallistel would call pilotage (compass orientation helped by familiar landmarks) Papi would say was not politage at all but map-based navigation. And what Papi would call map-based navigation, Baker would say was not true navigation at all but involved pilotage.

Such confusion over terminology may be irritating but it is not in the end very serious because just giving something a name like 'pilotage' or 'navigation' does not really explain anything at all. Even if we use such a term, we should only rest with it for a few seconds before going on to ask what exactly the animal is doing. If it is using familiar landmarks to find its way home, the important thing is to know how it is using them. Is it just moving from one beacon to another or does it attach a vector (distance and direction) to each landmark and so find its way home that way (Baker, 1984; Cheng, 1989)? Can it recognize the spatial relationships between several different landmarks and so get home even when placed at a new starting point? And so on.

In fact, one could argue that the different terminology adopted by different authors is actually a good thing because the definitions of each term are so variable that we are forced to say exactly what we mean and exactly what attributes an animal is supposed to have in a way that we might have got away with not doing if we could simply say 'pilotage' or 'map' and think we had explained anything at all.

In conclusion, many animals have cognitive maps in the sense that they have internal representations of geometrical relationships between objects in their environment. There are, however, a large number of different ways in which such internal representations can be organized and utilized by the animal. Some may involve quite simple recognition that a goal is 'below' or 'in the centre of' certain landmarks. Others may involve encoding a distance and direction with a series of landmarks within a familiar area (Baker, 1984; Collett, 1987; Cheng, 1989). Yet others may be more complex still. Just pin-pointing the cues (magnetic, visual, smell, etc.) an animal uses will not be enough. How animals use these cues to encode spatial information is what makes the study of cognitive maps truly informative and deeply fascinating.

Concept formation

Concept formation is another area of cognitive ethology where definitions have been troublesome and where, because of the clever-sounding nature of the term, people have sometimes been led to believe that animals are performing tasks that are more complex or more human-like than they really are (Herrnstein, 1990; Edwards & Honig, 1987; Medin & Smith, 1984; Roitblat & von Ferson, 1992).

The central question is, of course, what is meant by a concept. In everyday speech the word means an abstract category, one that is not tied to a particular type of physical stimulus or feature. Thus we probably would not use the word 'concept' to refer to the category of stimuli that give rise to our knee-jerk reflex or the category of yellow flowers because such categories can be directly related to physical features of stimuli in the world. But we would use it with 'the concept of the number three' or 'the concept of a free society' because there is no simple physical embodiment of all instances of what we are talking about.

In theory, it is abstract rather than physical properties that define a concept, but in practice such a distinction may be difficult to draw in a hard and fast way. 'Number' is clearly an abstract concept because it is not tied to any particular physical stimulus. The 'concept of the number three', for instance, can be applied to three pebbles of whatever size, colour or shape they are, three fish of any species, three blasts on a trumpet and so on. The concept of 'three' goes beyond (is abstracted from) particular physical stimulus configurations. At the other extreme, if all an animal could do was to discriminate tall vertical sticks from short horizontal ones, the term 'concept' would seem to be inapplicable. The animal might be able to categorize the world into cases where there were vertical lines and other cases where there were horizontal lines, but this could be related very directly to relatively simple physical features of the objects concerned. There would be minimal, if any, abstraction involved (Herrnstein, 1990).

While these two extremes of concept and non-concept are fairly clear, however, there are a large number of intermediate cases (as you might have guessed, the majority of actual animal examples that we have) which are much less easy to classify. Some people call them examples of concepts, thus implying that the animal is performing a high-level cognitive operation. Other people refer to the same examples as non-abstract responses to physical features. No wonder there is confusion as to what is really going on.

The root of the problem lies in the fact that it is, of course, extremely difficult to demonstrate that an animal is really responding to abstract properties and not to some much simpler physical property or properties that have escaped the notice of the experimenter. There are at least two ways in which animals can give the impression that they are using abstract concepts when in fact their response is based on physical features of their environment.

The first is when animals learn by rote that some stimuli should be responded to in one way and others in a different way. Such a task would correspond to those games where people are shown a trayful of objects for a few seconds and then get a prize for each one they are able to remember afterwards. The objects may have nothing in common at all except their being placed together on the tray. There is no abstract property that they all have in common – they just have to be remembered one by one.

Many animals, particularly birds, are outstandingly good at this sort of task. Vaughan & Green (1984) trained pigeons to discriminate 160 photographic slides randomly assigned to two groups. One group of slides was associated with a food reward, the other group was not, but there was no consistent difference between the two sets of slides – the pigeons simply had to learn by rote which was which. Not only could they do this with a high degree of accuracy, but they could do it just as well with a gap of 731 days with no further training.

This ability to remember long lists of stimuli is a particular hazard for anyone wanting to claim that an animal has formed an abstract concept because it may look as though the animal is learning something unrelated to the particular physical features of the stimuli. The temptation is then to conclude that it is an abstract property that has been learnt when in fact it may be no more than a list of physical features.

A good rote memory can operate even when the set of objects in question have little or nothing in common as far as obvious physical features are concerned. But when the objects share some, but not necessarily all, their physical features, there may be a second, even more hazardous trap for the unwary experimenter. This trap is called 'polymorphous features'.

Polymorphous features are those which members of a category set share in a probabilistic way. For example, not all members of the category 'house' look alike, nor do they all have windows, nor are they all made of bricks, nor do they all having sloping roofs. However, it would be possible to write down a list of physical features, such as parallel straight lines at each side, sloping lines at the top, etc. that define most houses. Not all houses have all these features but most would have some and a few might even have all of them. If we were to show that an animal such as a pigeon could discriminate between objects that we would describe as a house and those that we would not describe in this way, we might be tempted to conclude that the bird had developed an abstract concept of a house as 'a place where humans live' because it could recognize houses even when they appeared in such a wide variety of sizes, shapes and colours. However, if the pigeon had in fact learnt to recognize a house through using a set of polymorphous features (square and/or sloping roof and/or red brick and so on) such a conclusion would be quite unjustified. It could be using relatively simple physical features in a polymorphous or probabilistic way (Premack, 1983; von Ferson and Lea, 1990). There is no need – and indeed no justification – for invoking abstract concept formation in such a case.

To be quite sure that an animal has truly developed an abstract concept, then, it has to be shown that there is not a simpler basis for its categorization of the world, one based on physical features in either a simple or a rote learning or polymorphous way. Animal categorization of the world can become quite complex before we are entitled to say they are using abstract features that characterize true concept formation.

Are there any examples of concept formation in this strong sense? Roitblat and von Ferson (1992) suggest that one place we should search for it is in studies on so-called 'matching-to-sample' or 'matching-to-oddity' tasks where, by virtue of the way the experiment is set up, animals cannot be relying on simple cues. In a 'matching-to-sample' task, an animal is trained to choose one of two or more stimuli where the 'right' (i.e. food – rewarded) object changes from trial to trial depending on the 'sample' shown just beforehand. Thus, if a bird is learning to peck one coloured button among several different colours, it will only be rewarded for pecking the red key if a red 'sample' is shown just beforehand but only rewarded for pecking a green button if it sees a green sample button. Thus the button it has to peck to get food keeps changing and the animal cannot learn simply that red = food and green = no food. It has to learn that the rewarded key will be 'the same as' (in a matching-to-sample task) or 'different from' (in a matching-to-oddity task) what it is shown immediately beforehand. 'Same' and 'different' are abstract concepts, not tied to particular stimuli or features, and they argue that many different species seem to be able to operate on this level.

Another area that has been fruitful in indicating that non-human animals may form abstract concepts is the ability of animals to develop concepts of number (Davis & Pérusse, 1988). Some of the most convincing evidence comes from Pepperberg's (1987) study of an African grey parrot called Alex because in this case particular care was taken to eliminate all the other hazards we have discussed in this chapter. Reliance on simple physical features and rote learning, for instance, were carefully controlled for, and Alex can correctly say 'two', 'three' or 'four' to collections of objects he has never seen before. Even Pepperberg herself is not entirely satisfied, however, that he has developed a truly abstract concept of 'number' because it is always possible that he has utilized some polymorphous features of the training stimuli that are simpler than this. She is at present trying to see whether Alex can transfer his apparent concept of number learnt from visual stimuli to identifying numbers presented as sounds. If Alex can show evidence of seeing the similarity between three corks, three sticks and three noises, it will be difficult to conclude anything other than that he has an abstract concept of the 'number three'.

Conclusions: Do animals have cognitive representations?

It will be clear from the preceding discussion that we should be careful in applying the term 'concept formation' to any animal unless it can be shown that that animal is responding to abstract rather than physical properties in its categorization of the world. To do otherwise is to invite confusion even though in practice it may be difficult to be sure quite what categories an animal has.

Even if there is no evidence that an animal is forming an abstract concept, however, it can still be said to have a cognitive representation of the world. It is still legitimate to ask how a pigeon represents a tree or a house or another pigeon even though the way it does so may turn out to be through a list of polymorphous features and not anything more abstract.

We must remember, however, that even abstract representations may be quite simple. 'Same', 'different', 'the number three' may be abstract but they are hardly complex. The simplest computer program will involve evaluating whether a quantity is 'equal' or 'not equal' to another. All computers have an internal representation of numbers, too. To say that an animal has an abstract representation of the world, even an abstract representation, says very little about whether that representation is complex and nothing at all about whether that representation is conscious. A few lines of a computer language such as BASIC using '=' and '≠' could give a program that matched to sample or matched to oddity but it would do it very simply and it certainly would not do so consciously. A computer could recognize as 'three' a typed number, three times around a loop, the third item in a list and so on. All that is abstract is not conscious, in other words, and a considerable degree of confusion has been generated by a failure to realize this. What consciousness itself might be is the subject of the next chapter.

Further reading

Cheney & Seyfarth (1990) describe the representations that vervet monkeys might have while Heyes (1993) offers some interesting criticisms. Griffin (1993) and Byrne & Whiten (1988) provide food for thought.

9

Consciousness

Photograph by Marian Stamp Dawkins.

Until quite recently, it was considered unnecessary and, in some circles, unscientific for consciousness to be brought into discussions of animal behaviour at all. In both ethology and psychology there had been a long tradition that the only valid objects of study are objectively observable events – specifically, physiological changes happening inside an animal's body or behavioural events visible from the outside.

'Consciousness' with its inherent difficulties of observability (how can I ever know if you are conscious, let alone whether a non-human animal is?) was thought to fall outside the boundaries of legitimate science. Then, in 1976, Donald Griffin published an influential book called *The question of animal awareness* in which he questioned this total rejection of all attempts to study consciousness. He argued that consciousness was part of science and was, or should be, a legitimate part of animal behaviour studies. In this chapter, we will look at where the study of animal consciousness has got to since 1976 and at some of the difficulties that still beset anyone trying to study consciousness in animals. We will then venture into the difficult territory of animal 'suffering'.

What is consciousness?

One of the reasons that consciousness is so difficult to study is that an accurate comprehensive definition still eludes us. We use the word to cover so many different situations that a single definition may be impossible. We talk, for example, about being conscious of a toothache, of what time it is or of the vastness of the Universe. A feeling of intense pain, however, and a philosophical appreciation of the night sky share little in common beyond the fact that both give some sort of 'immediate awareness'. Both may occupy our minds and command our attention. This means that the most satisfactory or perhaps more accurately, least unsatisfactory definition of consciousness we can come up with at the moment is 'immediate, subjective awareness' (Griffin, 1993).

It is 'immediate' because it is a present sensation or experience. It is 'subjective' because it is private, known only to the being experiencing it. And 'awareness' expresses the range of phenomena that are usually included under the rubric of consciousness from pain to political opinions.

Even this crude definition is, however, enough to make it clear why 'cognitive' and 'conscious' are not the same thing. Cognitive, as we saw in the last chapter, refers to the information processing that goes on inside the brains of humans, non-humans and machines. But not all information processing, even in humans, is necessarily conscious (Velman, 1991). There are many tasks, such as calculating where a cricket ball will land, playing a musical instrument at speed or even recognizing objects, where much of the processing is unconscious and it is only in the final stages that something 'pops' into our consciousness. Memories may lie dormant for years until 'brought to mind'. We may even 'put things out of our minds' and find that, on returning to them, problems seem different or more tractable because of what has gone on unconsciously in the meantime. From Freud to more recent studies of brain-damaged people, it is recognized that much goes on in our minds that we are not consciously aware of.

This only serves to make consciousness even more mysterious. If so much can go on unconsciously, what do we need consciousness *for*? And – even more worrying from the point of view of trying to study consciousness scientifically – how do we know when information processing is conscious? With ourselves, we have the private 'inside' knowledge that we are sometimes aware of things and sometimes not. We know that we sometimes perform cognitive tasks such as finding our way around a route consciously and sometimes we do them without consciously thinking about what we are doing at all. But how could we know when this transition occurs for other people? Or other species of animal? Indeed does it occur at all? Might you not be the only conscious being in the world, with all the rest of behaving 'as if' we were but feeling and experiencing nothing?

The problem of consciousness

The previous section served to highlight the central problem of studying consciousness, the one that has proved so intractable that many people have concluded that any attempt to study consciousness must be hopelessly and irredeemably unscientific. This problem is that many, if not all the tasks we call 'cognitive' appear to be able to be carried out without the need for consciousness at all. Consequently, by simply looking at the behaviour (or brain physiology) of another person or animal, we would have no way of telling whether they were carrying out such tasks unconsciously or whether they were carrying them out with the added component of conscious awareness.

Certainly we could, as we discussed in the last chapter, discover how complex their cognitive representations of the world were, at least in theory, by doing the right experiments and seeing what behaviour resulted. An animal with a concept of number, for instance, behaves differently from one without such a concept. At least in principle we could devise an experiment to discover whether an animal had a concept of number.

Even in principle, however, no comparable experiments appear to be possible for detecting the presence of conscious awareness. There is no prediction we can make that if the animal has consciousness it should do X but if it is not conscious it should do Y. If an animal can consciously feel pain, for instance, we can predict that it will struggle and run away if we damage a limb or do something else to it that we know makes us feel pain. But this is not a unique prediction to consciousness. If the animal did not have conscious awareness of pain, it could still be struggling and squealing and doing its best to get away from the source of damage. Indeed natural selection would act to favour animals that removed themselves from danger and damage as quickly as possible. Nevertheless, as far as we can see, the struggling and squealing would be just as effective without the pain necessarily hurting the animal in the way it hurts us.

Similarly with clever or complex tasks, there is no unique prediction we can make to distinguish conscious from unconscious cognitive processing. The pigeon that can recognize stimuli as 'the same as' or 'different' may be doing it consciously, as we would probably do in many cases. But it may equally be no more conscious than a computer programmed to discriminate cases where two quantities were equal or not equal.

It is this inability to come up with unique predictions to distinguish conscious and unconscious cognitive processing or sensations that has given rise to the view that consciousness is unscientific. Respectable scientific theories have predictions that can be tested and then found to be true or false. Even hypotheses about adaptation, which have had enough accusations of 'Untestable!' hurled at them over the years are, as we saw in Chapter 1, vulnerable to being falsified by evidence and thus testable. But consciousness, that elusive sense of awareness of the world that we all know so well from the inside, appears to make no unique predictions that can be shown to be true or false. It might be occurring with any example of animal behaviour we care to name, but, equally, it might not be. It is untouchable, invulnerable to disproof and so unscientific. Or is it?

Functions of consciousness

There are basically two philosophical views about consciousness. One is that it plays no role whatsoever in the working of the brain, but is an extra, added on to the basic machinery, doing nothing except shedding its somewhat mysterious light over some of what the brain does. The other view is that consciousness is part and parcel of how the brain works, part of the mechanism in the sense that without it, the brain would function differently.

It is clear that if you hold the first view, known as epiphenomenalism, it does not make sense to ask what the function of consciousness is. If consciousness does not intervene in the workings of the brain, then it will be impervious to the action of natural selection. An animal or person that is conscious will be no better and no worse at doing anything at all than an animal that does the same task unconsciously. Natural selection cannot therefore favour the conscious over the unconscious animal and in this sense consciousness has no function.

Only if you hold the second view – that consciousness does something and makes a difference to how the brain operates – does it make sense to ask what the function of consciousness is. Natural selection, as we have seen throughout this book, needs differences between organisms to exist before it can get to work.

The problem is to identify what those differences might be. Why should a consciously aware animal be 'better' than one that 'goes through the motions' but does not have the light of conscious awareness? Is it more efficient? Quicker at arriving at solutions? More able to anticipate what other animals will do? Better able to plan ahead? Remember that if it were none of these things and conscious organisms were identical in all discoverable ways to unconscious automata, we should have to say that consciousness was an epiphenomenon after all and banish it from the scientific study of animal behaviour altogether.

Various functions for consciousness have now been suggested and, although many of them are ingenious, none of them is entirely satisfactory because none of them, it has to be said, is entirely free from the objection that the operations they are said to be involved in could not equally well be carried out without conscious awareness at all. For example, the 'Social Intellect' theory (Humphrey, 1983) holds that consciousness arose as a result of the complexities of social interactions between animals. The social environment of many animals including those which were ancestral to present-day humans, is or was full of problems such as which other individuals could be trusted and which could be successfully deceived. Such problems would favour the evolution of animals that were good at predicting the behaviour of others and 'second guessing' what they were likely to do, perhaps by imagining what they themselves would do in their place. While this is a powerful argument for complex cognitive representations of other animals, however, it is not clear why this demands a conscious rather than an unconscious representation.

Further clues as to the possible function of consciousness came from examining the situations in which humans are and are not conscious. Baars (1988) argues that many tasks can actually be performed better when we are not conscious of what we are doing than when we are. For example, playing a musical instrument can be smooth and effortless when a well-known piece is played unconsciously but much more difficult if the player thinks consciously about what each finger is supposed to be doing at a given time. Unconscious processors are often much better at dealing with predictable and often repeated events. Consciousness seems to come into its own when a novel situation is encountered and a completely new situation has to be worked out.

Perhaps, then, the function of consciousness is to solve problems of a sort that demand new lines of thinking, to rearrange the brain's view of the world or to enable it to work out before any action is taken just what the consequences of behaving in a new way might be. Many versions of these ideas have now been proposed but all of them leave a lingering doubt as to exactly why consciousness is necessary to deal with novelty. Simulations of the world have an obvious advantage in that, confronted with a new set of circumstances, an organism can work out what might happen in advance of trying out its various options in the real world, but it is not at all clear why this simulation could not be just as effective if it were done unconsciously.

In all this discussion of consciousness, however, we should not forget one important facet of what it is to be conscious. Certainly, we are often conscious when we think or reason, but so we are when we are in pain or thirsty. In other words, it is not just in complex cognitive processes that we have to explain consciousness but in basic sensations and feelings too. There is nothing particularly clever or novel about feeling pain. There may be very little thinking ahead beyond removing the source of pain. Equally, there seems nothing very cognitive about the sensation of seeing red or seeing blue, and yet we consciously experience sensations of colour. Clearly, any satisfactory explanation of the function of consciousness has got to be broad enough to include awareness of pain and colour as well as, say, working out what another organism or person might do in the future. It would also have to explain why removing ourselves from burning stoves with conscious feelings of pain, is more effective than going through the same action in a completely reflex way with no conscious awareness or sensation.

As yet, we have no theory comprehensive or big enough to explain convincingly the functions of consciousness in all these different contexts, despite the recent attempts that have been and are being made to do so. Dennett (1991) has argued that our whole view of what consciousness is is in need of radical revision and that we should stop thinking of it as a single stream. He sees it instead as a creative pandemonium in which different parts of the brain are simultaneously active and give different versions of the same events. Ornstein (1991) takes this idea even further and argues that consciousness is the transfer of dominance from one part of the brain with one set of actions to another part. For instance, if we consciously decide to suppress our anger at someone's remark and respond rationally and calmly, that would mean that a rational part of our brain assumes control over a more emotional part, whereas if we burst out with an impulsive insult, another part takes over and 'we' are angry.

Whether or not there proves to be anything in this idea of separate selves disciplined into a semblance of a unified 'self' by a shifting conscious awareness remains to be seen. It does, however, provide a working hypothesis in which we can see why some sort of control mechanism, conscious or not, becomes essential for humans and probably many other animals as well. If different parts of the brain were constantly competing for control of the motor system, there would be chaotic, disorganized behaviour with constant changes of 'mind'. Any animal that is capable of responding to a given situation in more than one way has to have mechanisms for preventing 'dithering' (McFarland & Houston, 1981), such as inhibition of all but one response or positive feedback to increase the chances that, once started, a behaviour will go through to completion. We know this happens at a reflex level quite unconsciously (Sherrington, 1906). We do not begin to pull our hand away from the stove and then wonder half-way through the movement whether there might be better ways we could be spending our time. The danger message takes precedence and, for the moment, inhibits other responses until it is complete. Similarly, movements to extend the limbs inhibit those to reflex the same limbs at the same time. Trying to do both sorts of movement at once would not result in 'creative pandemonium' but in chaos or complete immobility.

Sometimes, as with these examples of reflexes, order and discipline can be imposed on the behaviour of animals in a purely unconscious way. But at other times, the decisions about what behaviour to do next are not so clear-cut and the various options open have to be explored in a more dynamic way. For example, if a young red deer stag challenges an older male to a fight, neither may initially have sufficient information about the other's fighting ability to assess the likely outcome if a fight were to take place. However, after a period of mutual assessment, the challenger may withdraw without a fight or it may attack (Clutton-Brock & Albon, 1979). The 'decision' about whether to fight or not is based on information gained during the period of inspection.

We do not know whether this decision is taken consciously, as it might be if two human beings were 'deciding' whether to enter a fight, but it is clear that the representations that the stags have of each other and the rules they are following are a great deal more complex than reflexes to extend or flex limbs. It may be here, in the realm of decision-making and choices between different courses of action, that we should look for a role for consciousness, not necessarily, or at least not only, in the thinking, rational mind. I would like to end this chapter by applying this idea to one area of animal behaviour that is particularly controversial – namely, the measurement of animal suffering. It is a personal view of consciousness and so should not be taken as orthodox or standard theory. It does, however, emphasize the link between the emotional and the cognitive sides of awareness and so does bring together the many diverse phenomena we classify together under the single heading of 'consciousness'.

Animal Suffering

The word 'suffering' implies conscious awareness. If we did not believe that other animals experience pain and suffering (that their pain hurts, in other words) then we should be no more concerned about their suffering than we are about, say, damage to a statue or a Stradivarius violin. It is the belief that animals consciously experience pain and suffering in some of the same ways that we do that makes us talk about animal suffering rather than 'animal damage' or 'animal deterioration'. However, in view of the difficulties of knowing when animals are consciously aware of anything, how can we know animals are suffering?

'Suffering' when applied to humans is an interesting word because it covers so many different situations. We talk about suffering from hunger, from thirst, from boredom, from overwork, from a bereavement, from a war wound, from fear and so on. A huge range of external circumstances and bodily responses are included in this one word 'suffering', as are the very different conscious states that occur as a result. To experience grief is not the same as to experience thirst and yet something makes us group these states together. That something is that they are all aversive. They are all states that we would rather not be in and we will, if we can, make strenuous efforts to get out of them. If we are suffering from thirst, we will try desperately to find water, for instance. Suffering in humans denotes a state of being strongly motivated to bring about change either in internal state (such as water or food deficits) or external circumstances (such as imprisonment) and yet being unable to bring them about either because of physical thwarting or absence of suitable stimuli.

The same idea can be applied to other species. Animals, as we saw in Chapter 7, can be said to be motivated to do a whole variety of behaviour. A mildly thirsty animal in its normal environment will be motivated to drink and so will restore its fluid balance. Only if something very unusual happens – such as the animal being kept for long periods in a cage with no water – will the motivation to drink become so strong that the animal can be said to 'suffer' from thirst. 'Suffering', then, is not a supernumerary quality imposed on animals from the outside with no explanation. It is the activation of the normal mechanisms of motivation that animals have evolved by natural selection to optimize their fitness (Chapter 2), but where, for various reasons, the normal behaviour cannot be carried out. The animal may be motivated to run or flee, for instance, and be unable to do so in a small cage. It may be motivated to feed and there is no food. It may be motivated to dig a burrow and it has nothing but a concrete floor. So what we really need to do to study animal suffering is to measure the strength of an animal's motivation to do something. If the animal will literally 'do anything' to get something or get away from a certain situation, then we have a possible behavioural indication of suffering.

There are a number of different ways of measuring motivation, but one of the most promising in this context is to see whether animals will work to gain access to or to avoid something. Many animals, for example, will learn to press bars or peck keys to gain access to food and they can be made to 'work' even harder by arranging that they will get a pellet of food only if they press the bar or peck the key 5, 10, 50 or even 100 times. Not surprisingly, most animals that have been studied are prepared to work in this way and will keep on 'buying' the same amount of food even though they are having to 'pay' a lot more for each item. The animals are thus demonstrating by their behaviour that they are strongly motivated to obtain food.

Similar experiments can then be done with other rewards. Pigs that are about to give birth to piglets will build a nest, that is, if they are given access to straw or other suitable material. Most sows in commercial agriculture are, however, not given any nest material at all and the question can be asked as to whether they suffer as a result of this deprivation. Arey (1992) showed that not only would sows learn to push a panel to open a door that led to a room with straw, but that two days before giving birth they would push it up to 200 times just to be able to get to the straw and be able to make a nest with it. When Arey arranged it that the sows had to push one panel to go into a room with food and another to go into a second room with straw, the sows spent more time in the 'straw' room than the food room. The sows were showing by their behaviour that in the period before farrowing, straw was as important to them as food. They seemed to be so strongly motivated to obtain straw that it would be a reasonable conclusion to say that they would suffer (at least in the last few days of their pregnancy) if they do not have it or some similar material with which to make a nest.

One major advantage of such an approach to animal suffering is that it makes no assumptions about what would make a given animal suffer. It becomes an empirical question as to what leads to strong motivation in a given species or even a given individual of that species. We ourselves might not suffer if deprived of soil to dig in or straw to build a nest out of but other animals can show, by their behaviour, that they do suffer in such circumstances. It also means that we need make no assumptions about whether animals suffer if they cannot perform all the behaviour in their natural repertoire. It becomes a matter for empirical observation. A second advantage is that the measurement of animal motivation can be used in conjunction with other, more conventional measures of animal welfare and suffering, such as stereotypic behaviour and physiological disturbances, both of which have difficulties of interpretation if taken alone (Mason, 1991; Rushen, 1991). The animal's motivation brings us closer than anything else to the elusive conscious experience at the heart of what we mean by suffering.

However, as with other theories of consciousness, the idea that strong motivation to perform behaviour can be used as an indication of 'suffering' is also subject to the objection that it is impossible to know whether conscious experiences are accompanying the behaviour or not. We know that, in ourselves, if we were in a position to be giving top priority to getting water, say, and were so strongly motivated to drink that everything else was put aside in the search for water, we would be consciously suffering from thirst. The strong similarity between our behaviour in such circumstances and that of many other species suggests, but does not prove, that they too consciously experience states of suffering when they too are motivated to behave in certain ways. Consciousness retains its problematic status.

Conclusions

The problems of studying consciousness in any organism other than ourselves are immense and should not be underestimated. The view of many scientists that consciousness is not and never could be part of science has, however, recently been challenged and more and more people are now open to the possibility that at least some non-human animals are consciously aware of the world around them.

Apart from defining it, the two major interrelated problems with consciousness are (1) making predictions that are unique to conscious as opposed to unconscious organisms and (2) discovering a function for conscious awareness. Current theories of the evolution of consciousness generally fail to explain why the supposed advantages of being conscious could not be achieved just as successfully with unconscious processing of information.

The problem of studying the conscious experience of 'suffering' can be tackled by discovering how strongly motivated animals are to obtain or get away from something. Even this approach, however, relies on drawing an analogy between known conscious experiences in ourselves and assumed similar experiences in other species but only in a very restricted way. It is not assumed that animals are just like us in all their wants and desires but only in the specific respect of consciously experiencing pleasure when they get something they want and experiencing pain or suffering when in a situation they will work to avoid or get out of. We can then discover through experiment what those situations are. We make a leap between ourselves and other species. In the next chapter, we will turn the other way and ask whether the ideas that have taken up so much of this book on non-human animal behaviour can also be applied to our own species.

Further reading

Dennett (1991) gives an immensely entertaining account of his view of consciousness. Dawkins (1993) and Griffin (1993) weigh the evidence for consciousness in non-human animals. Bateson (1991) assesses pain in animals.

10

ANIMAL BEHAVIOUR AND HUMAN BEHAVIOUR

Homo sapiens and *Gorilla gorilla.* Photograph by Graham Wragg.

A human sciences undergraduate recently startled me by saying that she had been dreading doing the animal behaviour course because she had thought it was 'evil'. Fortunately, her view seemed to have modified as a result of actually doing some reading in the subject, but her remark did serve to emphasize the depth of the antagonism that many people have towards the idea of making comparisons between human behaviour and that of any other species. This antagonism comes from a number of different sources and is based on several distinctly different misconceptions, which will form the basis of this final chapter.

DETERMINISM OF VARIOUS SORTS

One of the main objections to comparing the behaviour of humans and non-humans – and the one most likely to give rise to accusations of the whole attempt being 'evil' – is that it implies a domination by genes, genetic determinism at its worst, in other words. If we are as we are because we have inherited our behaviour from our animal ancestors, then the fear is that there is nothing we can do about it. We are doomed, so this argument goes, to our genetic fate, with all our illusions of free will swept away by the force of our heredity. We have already dealt with the main fallacies in this argument in Chapter 4. There it was pointed out that many traits, particularly behavioural ones, may have some genetic basis but that does not mean that they will inevitably appear. Even something with clear physical symptoms and known genetic effects can be altered, modified and suppressed altogether by the right environment. People with the 'gene for' PKU do not necessarily develop PKU. Whether or not they do depends on what other genes they have and, in this particular case, what they eat. For a trait to have evolved by natural selection, it does not have to be completely specified by genes in the way that, say, blood groups are. All that natural selection requires is that there should be some genetic effects that appear under some circumstances. Environmental effects may still be massively more important in determining whether or not the trait actually appears in a given individual.

In fact, as we saw in Chapter 4, genes that influence behaviour are particularly likely to have their effects modified by the environment in which an animal finds itself. For instance, we saw that the 'genes for' maze learning ability in mice (meaning genetic differences between strains of mice) either were or were not apparent depending on what environment the mice were raised in. There was nothing inevitable about how quickly a given mouse learnt to run through a maze to find food, even though it might be genetically of the 'fast-learning' or the 'slow-learning' strain. For this reason, the suggestion that there might be 'genes for' human behavioural traits such as homosexuality or schizophrenia does not imply that anyone with certain genes will inevitably become gay or schizophrenic. If certain environmental conditions hold, they may, but equally they may go through life carrying the genes but being heterosexual and untroubled by mental illness.

Now, for some mysterious reason, people are less worried about the role the environment may play in determining behaviour than they are about genes. As Richard Dawkins put it, genetic determinism seems more determined than environmental determinism, despite the fact that the effects of early childhood, the trauma of accidents, diet and other experiences can be extremely influential in how people behave all their lives. Somehow people are more afraid of their genes than of the effects of what they have experienced, more likely to believe that they can 'break out' and exert their free will if they are told that an environmental factor is at work than if it is 'in their genes'. But think again of the example of the PKU children. Affected individuals can be helped by being given the correct diet – one free of phenylanaline. The effects of their genes can be largely overcome by altering their environment (in this case, food), but once an incorrect diet has wrought its effects and the person has developed symptoms of severe mental retardation, they cannot subsequently break out and become normal. The damage has been done. The early environment has had permanent effects.

The more we look, the more we see that we simply cannot support the claim that genetic effects are 'more determined' or less easy to modify than environmental ones. On the one hand, it is now possible to insert 'normal' genes into the bodies of people suffering from certain genetic diseases and to let the normal genes take over where their own defective ones would have failed them. Here genetic effects are overcome. On the other hand, it may be extremely hard, say, for drug addicts to give up their addiction. Here an environmental effect is difficult to overcome. So why should it be that some people are so uniquely afraid of genes and react as if the very idea of genes for behaviour meant monsters unleashed to dictate human affairs? Could it be that what they are really afraid of is the idea that their behaviour could be determined by *anything* and that to relate an aspect of human behaviour to a stretch of DNA or an extra chromosome seems to bring this idea uncomfortably closer to reality than some nebulous effect of a 'bad childhood' or a 'bad diet'?

This view is, however, not logical. It assumes that the way in which genes have their effects on behaviour is fundamentally different from the way in which the environment does and there is no evidence for this at all. Genes affect behaviour by altering some property of the physical body – hormone levels, for instance, or substances that affect the workings of nerve cells. But environmental effects can also be similarly biological in the sense that diet, temperature, daylength and so on can all affect the physical and chemical properties of our bodies such as hormone and neurotransmitter levels. Someone shouting at us, or having a near miss in a car accident can affect hormone secretion which will in turn affect mood and behaviour. Both are biological in the sense that some aspect of the physical body is changed and this change leads to a change in behaviour.

We cannot, then, escape from determinism by saying that we will reject genetic determinism because it affects the physical body and live with environmental determinism because it does not. Both do. Genes, childhood experiences, food – all affect our bodies and, through this, alter our behaviour. In the face of this, some people reject all kinds of determinism and say that science, however sophisticated its understanding of brains becomes, will never 'explain' human behaviour. They reject the scientific account of human beings altogether and say that there will always remain something extra, a spirit or free will, that can never be explained in terms of the workings of nerve cells and the secretions of hormones. There will be genes and environments affecting our behaviour but also a third freewill factor that cannot be attributed to either, but which just as surely guides what we do.

Other people reject the idea of a third factor and believe that what we do is a complex interaction between genes and environment. One day, they believe, science will understand how these two factors combine to produce the complexity of human behaviour.

Both of these views are logical in themselves but what would be illogical would be to attempt to hold a hybrid between these two views and claim that we were partly determined by genes and partly had free will because of the environment. The environment does not set us free from the way our bodies work. Psychiatrists, angry mothers, great literature not to mention more down to earth environmental effects like the length of winter days affect our bodies and their chemical make-up just as surely as genes do. We are the 'victims' of our environments just as much (and just as little) as we are the victims of our heredity.

So it is unfortunate that the arguments about genetic determinism are so often confused with the more general philosophical arguments about free will and determinism. Genes often take the full brunt of the arguments against determinism, with 'the environment' let off relatively lightly and indeed often incorrectly seen as an ally of free will. As we have seen, the effects of genes on behaviour are often highly modifiable: genes do not single-handedly determine behaviour. Environmental factors play a major part in whether genes are expressed or not. This much is a question of fact. What is a matter of opinion is whether genetic and environmental effects (which are both 'biological') give an adequate account of human behaviour. It is important not to confuse factual errors about the way genes work with the quite separate question of whether human beings can be said to have free will.

The importance of culture

A second argument against comparing human with non-human behaviour is that our culture is such a strong influence on what we do that it overrides everything else. Any genetic variation, that might well be demonstrated in another species, simply gets blotted out in us by the overwhelming influence of our cultural environment.

There are three answers to this bland dismissal of animal studies. The first is that non-human animals have cultures too. Rats transmit information about what foods are poisonous from generation to generation, many birds learn what predators look like and what to do about them from their parents and chimpanzees even learn from each other how to use stones to crack open nuts. These are examples of 'culture' in the sense that a skill or information possessed by one animal is passed from one individual to another and down the generations by non-genetic means (Bonner, 1980; McGrew, 1992). But to critics of the human–non-human comparison, this is not culture in the human sense, which means music, religion and literature and altogether higher and finer things than just learning what to eat or what is dangerous. Non-human animals, on this view, do not have cultures, whereas humans do. But if anyone adopts this manœuvre, we can bring out a second reply, one that is much more difficult to dismiss than a simple appeal to definition.

This second reply is that, despite the undoubted importance of culture in our lives, there are nevertheless parallels between our behaviour and that of non-human species. We do do things that are predicted by the theory of natural selection, despite our wanting to think that we have escaped from all that and put ourselves a long way away from the rest of the animal kingdom by virtue of our uniquely complex 'culture'. If we really take the trouble to look, we find that the distance may not be as great as we would like to think, even in something as intimate as our choice of marriage partners.

We have already seen that mating among non-human animals is far from random. Females, in particular, are choosy about who they mate with, and many males develop elaborate ornamentation under the selection pressure to be the one chosen. It is clearly important for a female to choose a healthy male of the right species but, in addition, it is also important for her to avoid too much inbreeding. Mating with close relatives (parents, offspring or siblings) leads to inbreeding depression, that is, to smaller numbers or less healthy offspring (Cavalli-Sforza & Bodmer, 1971), due to the high chance that damaging recessive genes carried by one partner will also be present in the other if it is a relative.

Natural selection, therefore, should favour animals that avoid mating with close relatives and indeed we find that incestuous (parent–offspring or sib–sib) matings are very rare in nature – less than 2 per cent in birds and mammals (Harvey and Ralls, 1986). Animals avoid incestuous matings either because one sex moves away from the natal area and finds a mate elsewhere (as happens in male Gelada baboons and many female birds) or the siblings of both sexes remain in the same general area and develop an aversion to mating with each other. They become more attracted to unfamiliar (and probably less related) individuals.

Pusey (1980) describes in detail the change in the behaviour of young female chimpanzees as they become sexually mature. Before sexual maturity, they associate freely with the young males in their troop and may even have a favourite male companion, often their brother. But once their oestrous cycles begin and they are able to reproduce, they no longer associate with this favourite male, and only very rarely copulate with him. In fact, they may travel quite long distances and visit another troop in order to mate, despite the fact that the males in their own troop are much closer. The very familiarity of their own males seems to make them unattractive, but the net result is avoidance of inbreeding.

In humans, there seems to be a similar avoidance of mating with familiar siblings. Shepher (1971) found that people brought up in the same peer group in an Israeli kibbutz did not marry each other, despite having a great deal in common and all being of a suitable age. The experience of having been brought up together as 'pseudo' brothers and sisters seemed to have destroyed sexual attractiveness between them. Usually, individuals brought up together are related, so a tendency to avoid marrying a familiar individual could be seen as an adaptation to avoid inbreeding, a characteristic shared by humans and non-humans (van den Burghe, 1983), and shaped in both cases by natural selection. Not everyone will accept that this aspect of mate choice in humans has been subject to the same evolutionary pressures that have undoubtedly been at work in other species. They will prefer to see it as due to culture, rather than genetics, to taboo not a way of raising fitness. But the parallels, both in outcome (avoiding inbreeding) and in mechanism (not mating with familiar individuals) are undoubtedly there. Despite or even because of our culture, we behave remarkably like other species.

In other aspects of human mate choice, too, parallels, possibly uncomfortable ones, exist. Human females consistently rank such traits as ambition and financial prospects higher than males in the qualities sought for in a prospective partner. Human males, on the other hand, rank age and physical qualities related to ability to have children higher than do females (Buss, 1989). Of course it would be possible to argue that these differences are purely cultural and have nothing to do with fitness but they do seem to hold across many different cultures and, what is more, they are consistent with human mate choice being an evolutionary adaptation.

On the hypothesis that human mate preferences have evolved under natural selection, we would expect there to be consistent differences between the sexes. In humans, child-bearing is immensely more burdensome than the act of copulation and so female reproductive prospects are much more dependent on age and health than are male ones. This means that males should be attracted to partners that are young and able to cope with the rigours of pregnancy. Females, on the other hand, should show a different pattern of choice. Females in an environment where food and other resources were scarce, would gain reproductively from having a wealthy, high-status partner who would be able to give material benefits to them and their offspring. Females would therefore be expected to be attracted to men who are rich and of high status.

This difference in the attributes of a preferred mate for human males and females is exactly what has been shown by extensive studies using nearly 10,000 people in 37 cultures spread over six continents and five islands (Buss, 1989, 1992; Ellis, 1992). Males are more influenced by the age and physical attributes of potential mates, whereas females rate achievement and status more highly. Furthermore, Pérusse (1993) found that amongst French Canadians, social status in males (as judged by income, education and so on) did not correlate with the number of children they had but did correlate well with numbers of copulations, which, in the absence of contraception, would have been a good measure of fitness and female preference. So even in a modern society, where fitness is thwarted by contraception, then, we can see patterns of human mating behaviour that are consistent with an adaptive interpretation. Despite culture, despite diversity, human mate preferences do seem remarkably consistent with what would raise reproductive success.

But the die-hard critic might still say that these are minor factors in the grand scheme of human affairs. Although there might be some parallels to be drawn in some small aspects of sexual attraction and even other aspects of human life, culture is still the dominant factor in human behaviour. Look at the diversity of human beings, he or she might say, and the extraordinary feats of technology that humans have achieved, far beyond anything that any animal, even a chimpanzee with a stone tool, might aspire to. Look at Chartres

Cathedral, look at the Pyramids, look at the Aztecs and the centuries of Chinese civilization.

The danger of pushing this argument too far is that it soon runs headlong into the sort of difficulty that its proponents least expect to be troubled by: the biological basis of culture. The more the achievements of human culture are trumpeted, the more difficult it becomes to maintain the dichotomy of culture *or* genetics. Culture is not some extra-terrestrial force that arrived to take over organisms that were up to that moment totally in the grip of their genes. Culture, or rather the tendency of individuals to respond to cultural influences, is rooted in biology. We are cultural beings because we have been shaped by natural selection to be so.

For example, the language we speak is clearly determined by which culture we grow up in, but the capacity for speaking a language at all and even which stage of our lives we are best at learning it, is plausibly seen as a biological adaptation (Pinker & Bloom, 1992). Thus culture determines whether we speak Japanese or Hauser, but our capacity for learning and using language depends on our brain structure and vocal apparatus. Genetics gives us the basic tools, although quite how we use them will of course be profoundly influenced by what we experience in the course of our lives. The genetic programs are open-ended, designed to be influenced by the specific environment in which a baby human finds itself, such as the language it hears. Culture cannot escape from biology.

The same is increasingly being seen to be true of many other aspects of human culture (Williams, 1992; Symons, 1992). Our tendency to be cultural at all, to take on the norms and customs of the societies we grow up in, to defend our own group or traditions – these are such universal characteristics of human beings that we can suspect them of having been strongly selected for in the past. Certainly, in the examples of 'culture' (or, if you prefer, 'proto-culture') seen in non-human animals, there are clearly biological advantages in one animal receiving information from another and doing what it does. A young rat that eats where its parents eat and learns to take the food that they do is more likely to eat safe food than a young rat that tries everything for itself including rat poison. A young chimpanzee that watches its mother catch termites by pushing a twig into a termite mound and letting the insects crawl up it is more likely to be able to eat termites itself than one left totally to its own devices and the off-chance that it would 'reinvent' the art of termite fishing. The capacity to learn for oneself, in other words, can be a mixed blessing. On the one hand it gives flexibility and an ability to keep up with a changing environment. On the other hand, the 'error' part of 'trial-and-error' can be extremely time-consuming and often dangerous. But by learning from another animal, there can be the best of both worlds. There are all the advantages of new, complex adaptive behaviour, without the disadvantages of having to taste the poisonous food and nearly die or waste hours of time in trying to discover something that another animal has already perfected.

This raises an important point. Natural selection, in these cases acting to favour young animals whose feeding behaviour is influenced by that of their parents, will lead to a diversity of results. Rats in different areas will avoid different foods. Chimpanzees from different geographical locations will adopt different feeding techniques. Evolved adaptations, in other words, will not necessarily be identical in all members of a species. The argument, common among social scientists, that human behaviour cannot have evolved by natural selection because it is so diverse, is based on a misconception of the way natural selection works (Williams, 1992). Genetic selection, acting on a learning mechanism or giving different results in different circumstances will lead to diversity, not species-specific uniformity.

This is going to be particularly true of our own species. We, as individuals, would be far worse off if we had had to single-handedly reinvent the wheel, the dynamo or the bridge. We have gained enormously from the cumulated wisdom and tradition of other humans and in so doing, we have become very diverse in our habits and customs. So do not think that by simply waving the banner of 'culture' you have got rid of the case for comparing the behaviour of human and non-human animals. To many biologists, 'culture', even in humans, is yet another product of natural selection, a means of survival, a way of being able to reproduce in a hostile world. A means of survival?, you might say. Is not our culture, with its destructive tendencies and its capacity for killing and destroying the environment going to make it difficult for any of us to survive? Does this not show that culture and genes are in opposition? Does it not therefore prove that culture cannot come from genes and that we are different from all other species in having the ultimate in non-adaptive behaviour?

Time lags

Natural selection is short-sighted. Fitness advantages are measured, as we saw earlier, largely in terms of numbers of children and grandchildren. If a gene is eliminated from the population, that is the end of it unless it reappears because of a new mutation. There is no one around to say: it may be disadvantageous now but it will be helpful in a few million years, so keep it. Equally, there is no one around to say: well, this looks like a useful mutation, but you wait, it will have disastrous effects in a few thousand generations, so get rid of it now. If there is a short-term advantage of one genotype over another, that genotype will prevail, and if a longer-term disaster happens as a result, there is nothing to stop it. The disaster will happen.

Of all the difficulties of comparing human and non-human behaviour, the most difficult to deal with is this issue of the conflict between short-term and long-term advantage, or to put it another way, the time lag between when human behaviour evolved and the present day. It was already apparent in Chapter 1, that many animals may behave non-adaptively because their environments are changing too rapidly. They are adapted to an environment of the past and have not yet 'caught up' genetically. We could use many examples, particularly where humans have imposed massive recent change on an animal's habitat. Toads migrating back to the ponds where they were spawned and then getting squashed in large numbers by cars on the way, is an obvious one, so obvious in fact that we hardly think it strange. It is not necessary to spend a great deal of time worrying about the adaptive significance of being squashed on roads. It is obvious that toads were selected to move to ponds in the breeding season and that cars are a new development for which natural selection has not (yet) equipped them.

With humans, such time lags are even more obvious and may only now be catching up with us. Our ancestors spent the last 2 million years as Pleistocene hunter-gatherers. The few thousand years that have elapsed since the rise of agriculture in different parts of the world represent less than 1 per cent of this time and there is no evidence at all that selection has acted differently on people who adopted agriculture and those that remained with the hunter-gatherer way of life (Tooby & Cosmides, 1992).

In other words, human behaviour now is the legacy of that long period as hunter-gatherers, when the world was very different. Our environment has changed, often by our own hands and we are left with behavioural traits that were highly adaptive then but may be positively damaging now. A tendency to like sweet food, for example, could have evolved at a time when the sweetest things around were ripe fruit and the occasional cache of honey. Eating anything sweet that was discovered could have been highly beneficial in a way that in today's environment, with its readily available confectionery, it simply is not. Instead of puzzling about why we should like something that is bad for us, we have to cast our minds back to the situation that probably existed in the Pleistocene when such taste preferences evolved and ask what the advantage of it was then. And that is where our problem lies. When did the behaviour evolve and what was the situation like then?

Human behaviour may have evolved, but it has time lags of such monumental proportions that showing its adaptive nature in the way we discussed for non-human animals in Chapter 1 becomes much more difficult. For most other species, we can either observe them in their natural environments or have a fairly clear idea as to what their natural environments are or were. With our own species, we have so altered our environment, by our own efforts, that it is difficult to know where to begin.

Nevertheless, it is clear that we should make some effort to do so. The question 'What is the advantage of building an atomic bomb?' is not a sensible question. The human inventiveness that gave rise to this destructive weapon evolved long before that inventiveness was put to this particular use. The original selective advantage of inventiveness was probably to develop tools or to devise solutions to practical problems. If inventive proto-humans did better than less intelligent ones, there would be nothing to stop them thriving and reproducing and giving rise, many generations later, to inventive descendants who would develop the atomic bomb. The question should be 'What was the advantage of inventiveness?' because it was that that natural selection got to work on all those years ago.

Sometimes, as we saw in the last section, there is a present-day advantage to what humans do. Marrying a close relative does have immediate genetic consequences. High-status men do have more copulation opportunities, even though this may not translate directly into numbers of children because of contraceptives. Under such circumstances, the adaptive nature of human behaviour emerges despite the effects of culture and time lags. But much of what we do seems anomalous, even maladaptive and here it is right to be much more cautious. It would be a mistake, however, to assume immediately that therefore the behaviour has not evolved by natural selection because what we do now gives no fitness advantages. It may be that we are looking at a trait that once was adaptive but is now no longer so. The environment may have been so changed by human activity that its original adaptive advantage is totally obscure. A great deal of caution is therefore needed in interpreting the adaptive nature of human behaviour in the face of what might be called the biggest evolutionary time lag of all time.

Which animals?

The final challenge that can be made to the idea of comparing the behaviour of human and non-human animals is over which animals should be used for the comparison. There are, to put it mildly, rather a lot of them and many of them (racoons and stick insects, say) differ radically from each other. So comparing humans with other species would seem to be quite pointless since the answers we get will be different depending on which species we choose. Even among our closest animal relatives, the apes, we find a great deal of diversity. Chimpanzees live in groups and have a promiscuous mating system, gibbons are monogamous and live in pairs with their offspring, gorillas travel in groups with one male with several females and orang-utans lead largely solitary lives apart from the necessities of copulation and child care. Which one should we compare ourselves with?

I would like to argue in this final section that the answer is none, or rather, all. Each species is unique. Even ten-spined sticklebacks differ quite considerably from three-spined sticklebacks and humans differ radically from even their closest relatives, the great apes. When comparisons are made between non-human species, they become most valuable when they do not involve just two species but large numbers, all following a similar way of life. In Chapter 1, we saw that the Comparative Method, that studies adaptation by comparing species, comes into its own when it can give an explanation for behaviour in phylogenetically very distinct animals. Comparing eggshell removal in the black-headed gull with its absence in the closely related kittiwake is suggestive of the influence of nest predation on behaviour because black-headed gulls nest on the ground where predators are a threat and kittiwakes nest on cliffs which predators cannot reach.

But the evidence from just two species is not very conclusive. As we saw earlier, it becomes much stronger when we compare many different birds that rely on camouflage to protect their nests and do remove eggshells (such as nightjars) with a wide range of birds that do not rely on camouflage and also do not remove eggshells. It is the independent evolution of the trait (shell removal) in phylogenetically distinct birds all subject to the same dangers of nest predation that provides some of the most convincing evidence that shell removal is an adaptive response to hide the nest from predators.

Similarly, if we are looking for an adaptive explanation for human behaviour, we should not just look at one other species, but see whether we can see parallels in a wide range of other species. If it were just chimpanzees and no other animal where females avoided mating with their brothers, we could probably not conclude very much about inbreeding avoidance in humans. But the fact that avoidance of inbreeding, either by dispersal of one sex or positive preference for unfamiliar mates, is such a widespread phenomenon in birds and mammals (Clutton-Brock, 1991), makes the existence of an adaptive inbreeding-avoidance mechanism in humans much more plausible. In fact, the similarity between the way in which humans and other animals avoid mating with individuals with whom they have grown up puts the onus on the critic to explain why the similarity exists if natural selection has not been at work in similar ways in both cases.

What we should look for in comparing humans with other species, then, are generalizations that already hold true for large numbers of non-human species. Single comparisons between humans and one other selected species, however closely related, could be extremely misleading as so much would depend on which other species were selected. For example, if we look at mammals as a whole, we find that mothers that suckle their young infrequently have much more concentrated milk than those that suckle them much more often (Ben Shaul, 1962). One of the most extreme examples of concentrated-milk species are tree-shrews, where the young are placed in a separate nursery nest. The mother visits them only once every 48 hours and gives them a meal of extremely high fat, high protein milk (Martin, 1984). In many other primates and marsupials, on the other hand, where the young are in virtually constant contact with the mother, the milk is much more dilute and delivered much more frequently. Given that this is a pattern that holds over a large number of species, we can then ask where humans fit into this overall mammalian pattern. It turns out that we are in the relatively weak milk category, along with other species that have relatively helpless young and constant interaction between mother and offspring. We thus do not compare ourselves with one other species but many. We do not ask whether we are like gorillas or like fur seals or like tree shrews but whether we fit into a more general pattern of mammalian variation.

The stronger that pattern is among diverse non-human species, the more confident we can be that it is at least worth asking whether humans do or do not fit into the same pattern. If we do not, then that is evidence for saying that humans are different. Perhaps culture or some unique feature of our way of life has made us the exception. That would be an entirely legitimate conclusion if that is what the data pointed to. But if we do fit into the general pattern, then that is evidence for saying that similar selective processes have been at work on us as have been identified for other species. The point is that the acceptance or rejection of a particular selective effect on human behaviour is an empirical matter, one that should be based on facts about humans and not just arising out of preconceived convictions that human behaviour obviously is (or, depending on your point of view, obviously not) touched by natural selection in at least some of the same ways that behaviour and structure in other species are.

For this reason, too, it is important to look not just at other primates as our closest living relatives, but at a wide range of species that may be phylogenetically more distant from us but may share as many, if not more, characteristics. Among mammals, for example, monogamy is relatively rare and polyandry (one female with several males) unknown. But among birds, monogamy is common and polyandry is also known to occur. So if we want to understand the range of human mating systems, which includes monogamy, polygamy and polyandry, it might be just as important to study birds and understand why they have such a wide range of mating systems as to study chimpanzees and gorillas (Wilson and Daly, 1992). In fact, some birds such as the dunnock (*Prunella modularis*) have as much within-species variation in their mating systems as do humans and can be monogamous, polygynous, polyandrous and even polygynandrous with both sexes sharing mates (Davies, 1992). This is not to say that humans are 'like dunnocks' in the sense that these particular birds provide a model of human mating patterns. Rather, they illustrate the fact that a genetic adaptation – in this case a species' mating system – can take many forms and lead to great diversity. They also underline the main point of this section, namely, that to understand our own behaviour we may have to look across a wide range of species and to include animals that at first sight seem to be very different from ourselves.

Conclusions

The purpose of this chapter has been to argue that certain common objections to the idea of comparing the behaviour of humans and non-humans do not hold water. The objections to 'genetic determinism' turn out to be based on misconceptions about genetics, or worse, a blanket objection to *any* explanation of human behaviour. The objection that culture is the only important factor for humans turns out to be factually incorrect and based on an erroneous dichotomy between cultural and genetic adaptation. The problem of time lags in human evolution is serious and means that apparently maladaptive human traits have to be interpreted with caution. The problem of which human–animal comparisons are most valid is best resolved by using only comparisons that have already been shown to be based on good correlations among many species of non-humans.

Drawing parallels between the behaviour of humans and other species is fascinating, controversial and fraught with difficulties. It is my hope that this chapter will have given a glimpse of some of the fascination, shown a way through some of the controversies and shown that at least some of the difficulties are not quite as formidable as they are sometimes made out to be.

FURTHER READING

The articles by Tooby & Cosmides (1992) and Symons (1992) should be compulsory reading for anyone interested in the application of Darwinian ideas to human behaviour. Durham (1991) discusses the evolution of culture.

EPILOGUE

There may be some people who will feel that seeing animal behaviour (particularly human behaviour) as a means for passing genes on from one generation to the next somehow belittles or diminishes it. Perhaps they will feel that all the things we have been discussing in this book – caring for relatives, choosing a mate, fighting and so on – have been made to seem 'nothing but' the working out of this process. I would like to end on a personal note by saying that, for me, nothing could be further from the truth. To give an explanation of something may take away its mystery but it does not take away our capacity to marvel at it. In fact, it may become more marvellous when we realize what has given rise to it.

It does not diminish Abraham Lincoln's achievements to say that he was 'nothing but' a backwoodsman. We admire him and what he did the more when we realize that he did not start from a position of privilege but that what he did came from his own efforts and personality. His humble beginnings actually make his final achievements more remarkable. So, in an infinitely greater way, the simple beginnings of animal behaviour – the battle of the shifting gene frequencies – make the complexity of what we see animals doing into something truly remarkable.

The fight for a place in the gene pool has given rise to some of the most beautiful and intricate phenomena on this earth. Animals are not formless masses of jelly, mindlessly reproducing themselves. They have developed the power to swim and to fly, to care for their young, to stalk their prey, to play, to sing and be curious about the world around them. Some of those powers, plus some additional ones, we see in our own species. To know that all this comes from such simple beginnings can enhance and deepen the wonder, not diminish it.

The animal that builds a nest and cares for its young may be 'nothing but' the result of natural selection, but what a result! From the action of selection, favouring one kind of animal over another down millions of years, have come cooperation, lifelong bonds between individuals and, ultimately, the ability to reason and understand it all. I can think of few things as marvellous or more worthy of a lifetime's study than that.

CHESTER COLLEGE LIBRARY

REFERENCES

Abraham, M. V. (1993) The trade-off between foraging and courting in male guppies. *Animal Behaviour* **45**: 673–681.

Adams, E. S. & Caldwell, R. L. (1990) Deceptive communication in asymmetric fights of the stomatopod crustacean, *Gonodactylus bredini*. *Animal Behaviour* **39**: 706–716.

Alatalo, R., Lundberg, A. & Glynn, C. (1986) Female pied flycatchers choose territory quality not male characteristics. *Nature Lond.* **323**: 152–3.

Altman, J. S. & Kein, J. (1990) Highlighting *Aplysia's* networks. *Trends in NeuroSciences* **13**: 81–82.

Andersson, M. (1982) Sexual selection, natural selection and quality advertisement. *Biological Journal of the Linnean Society* **17**: 375–393.

Arak, A. (1988) Female mate selection in the natterjack toad: active choice or passive attraction? *Behavioural Ecology and Sociobiology* **22**: 317–327.

Arey, D. S. (1992) Straw and food as reinforcers for prepartal sows. *Applied Animal Behaviour Science* **33**: 217–226.

Austad, S. N. & Rabenold, K. N. (1985) Reproductive enhancement by helpers and an experimental examination of its mechanism in the bicolored wren: a facultatively communal breeder. *Behavioural Ecology and Sociobiology* **17**: 19–27.

Baars, B. (1988) *A cognitive theory of consciousness*. Cambridge: Cambridge University Press.

Baker, R. R. (1984) *Bird navigation: the solution of a mystery?* London: Hodder & Stoughton.

Balmford, A., Thomas, A. L. R. & Jones, I.L. (1993) Aerodynamics and the evolution of long tails in birds. *Nature Lond.* **361**: 628–631.

Barnard, C. J. (1991) Kinship and social behaviour: the trouble with relatives. *Trends in Ecology and Evolution* **6**(11): 310–312.

Bateson, P. P. G. (1983) *Mate choice*. Cambridge: Cambridge University Press.

Bateson, P. P. G. (1991) The assessment of pain in animals. *Animal Behaviour* **42**: 827–839.

Bateson, P. P. G. (1993) The need to promote behavioural biology. *ASAB Newsletter* **19**: 9–11.

Bekoff, M. (1992) Scientific ideology, animal consciousness and animal protection: a principled plea for unabashed common sense. *New Ideas in Psychology* **10**: 79–94.

Bell, G. (1982) *The masterpiece of nature*. London: Croom Helm.

Ben Shaul, D. M. (1962) The composition of the milk of wild animals. *International Zoo Yearbook* **4**: 333–342.
Bentley, D. & Hoy, R. (1972) Genetic control of the neuronal network generating cricket (*Teleogryllus*) song patterns. *Animal Behaviour* **20**: 478–492.
Bonner, J. T. (1980) *The evolution of culture in animals.* Princeton: Princeton University Press.
Borgia, G. (1985) Bower quality, number of decorations and mating success of male satin bowerbirds (*Ptilinorynchus violaceus*): an experimental analysis. *Animal Behaviour* **33**: 266–271.
Burrows, M. & Rowell, C. H. F. (1973) Connections between descending interneurons and metathoracic motorneurons in the locust. *Journal of Comparative Physiology* **85**: 221–234.
Buss, D. M. (1989) Sex differences in human mate selection criteria: evolutionary hypotheses tested in 37 cultures. *Behavioral and Brain Sciences* **12**: 1–14.
Buss, D. M. (1992) Mate preference mechanisms: consequences for partner choice and intrasexual competition. In: *The adapted mind* (eds. J. H. Barkow, L. Cosmides and J. Tooby) New York: Oxford University Press, pp. 249–266.
Byrne, R. W. & Whiten, A. (1988) *Machiavellian intelligence, social expertise and the evolution of intellect in monkeys, apes and humans.* Oxford: Clarendon Press.
Cade, W. (1981) Alternative mating strategies: genetic differences in crickets. *Science* **212**: 563–564.
Caraco, T., Blanckenhorn, W. U., Gregory, G. M., Newman, J. A., Recer, G. M. & Zwicker, S. M. (1990) Risk sensitivity: ambient temperature affects foraging choice. *Animal Behaviour* **39**: 338–345.
Cavalli-Sforza, L. L. & Bodmer, W. F. (1971) *Genetics of human populations.* San Francisco: W. H. Freeman.
Cheney, D. L. & Seyfarth, R. M. (1988) Assessment of meaning and the detection of unreliable signals by vervet monkeys. *Animal Behaviour* **36**: 477–486.
Cheney, D. L. & Seyfarth, R. M. (1990) *How monkeys see the world.* Chicago: Chicago University Press.
Cheng, K. (1989) The vector sum, model of pigeon landmark use. *Journal of Comparative Psychology* **A166**: 857–863.
Clark, D. L. & Uetz, G. W. (1992) Morph-independent mate selection in a dimorphic jumping spider: demonstration of movement bias in female choice using video-controlled courtship behaviour. *Animal Behaviour* **43**: 247–254.
Clutton-Brock, T. H. (1991) *The evolution of parental care.* Princeton University Press.
Clutton-Brock, T. H. & Albon, S. D. (1979) The roaring of red deer and the evolution of honest advertisement. *Behaviour* **69**: 145–170.
Clutton-Brock, T. H., Guiness, F. E. & Albon, S. D. (1982) *Red deer. Behaviour and ecology of two sexes.* Edinburgh: Edinburgh University Press.
Colgan, P. (1989) *Animal motivation.* London: Chapman & Hall.
Collett, T. S. (1987) Insect maps. *Trends in Neuroscience* **10**: 139–141.
Coopersmith, C. B. & Lenington, S. (1991) Female preference based on male quality in house mice: interaction between male dominance rank and t-complex genotype. *Ethology* **90**: 1–16.
Cowie, R. (1977) Optimal foraging in great tits, *Parus major*. *Nature, Lond.*, **268**: 137–139.

Cullen, E. (1957) Adaptations in the kittiwake to cliff-nesting. *Ibis* **99**: 275–302.

Cullen, J. M. (1960) Some adaptations in the nesting behaviour of terns. *Proceedings of the XIIth International Ornithological Congress, Helsinki*, pp. 153–157.

Cullen, J. M. (1972) Some principles of animal communication. In: *Non-verbal communication* (ed. R. A. Hinde). Cambridge: Cambridge University Press, pp. 101–122.

Darwin, C. (1871) *The descent of man and selection in relation to sex*. London: Murray.

Davies, N. B. (1992) *Dunnock behaviour and social evolution*. Oxford: Oxford University Press.

Davies, N. B. & Halliday, T. R. (1978) Deep croaks and fighting assessment in toads *Bufo bufo*. *Nature Lond.*, **274**: 683–685.

Davis, H. & Pérusse, R. (1988) Numerical competence in animals: definitional issues, current evidence and a new research agenda. *Behavioral and Brain Sciences* **11**: 561–615.

Davis, J. M. (1980) The coordinated aerobatics of dunlin flocks. *Animal Behaviour* **28**: 668–673.

Dawkins, M. S. (1993) *Through our eyes only? The search for animal consciousness*. Oxford: W. H.Freeman.

Dawkins, R. (1979) Twelve misunderstandings of kin selection. *Zeitschrift für Tierpsychologie* **51**: 184–200.

Dawkins, R. (1982) *The extended phenotype*. Oxford: W. H. Freeman.

Dawkins, R. (1986) *The blind watchmaker*. London: Longman.

Dawkins, R. & Krebs, J. R. (1978) Animal signals: information or manipulation? In: *Behavioural ecology*, 1st edn. (eds. J. R. Krebs & N. B. Davies). Oxford: Blackwell Scientific Publications. pp. 282–309.

Dennett, D. C. (1991) *Consciousness explained*. Boston: Little, Brown and Co.

Drent, R. H. & Daan, S. (1980) The prudent parent: energetic adjustments in avian breeding. *Ardea* **68**: 225–252.

Durham, W. H. (1991) *Coevolution: genes, culture and human diversity*. Stanford: Stanford University Press.

Edwards, C. A. & Honig, W. K.(1987) Memorization and 'feature detection' in the acquisition of natural concepts in pigeons. *Learning and Motivation* **18**: 235–260.

Elgar, M. A. (1986a) House sparrows establish foraging flocks by giving chirrup calls if the resources are divisible. *Animal Behaviour* **34**: 169–174.

Elgar, M. A. (1986b) The establishment of foraging flocks in house sparrows: risk of predation and daily temperature. *Behavioural Ecology and Sociobiology* **19**: 433–438.

Elgar, M. A. (1989) Predator vigilance and group size among mammals: a critical review of the evidence. *Biological Reviews* **64**: 13–34.

Ellis, B. J. (1992) The evolution of sexual attraction: evaluative mechanisms in women. In: T*he adapted mind* (eds. J. H. Barkow, L. Cosmides & J. Tooby). New York: Oxford University Press, pp. 267–288.

Emlen, S. T. & Wrege, P. H. (1989) A test of alternate hypotheses for helping behaviour in white-fronted bee-eaters of Kenya. *Behavioural Ecology and Sociobiology* **25**: 303–320.

Endler, J. A. (1980) Natural selection on colour patterns in *Poecilia reticulata*. *Evolution* 34: 76-91.

Endler, J. A. (1986) *Natural selection in the wild.* Princeton: Princeton University Press.

Endler, J. A. (1993) Some general comments on the evolution and design of animal communication systems. *Philosophical Transactions of the Royal Society of London* **B 340**: 227–230.

Ewert, J.-P. (1987) Neuroethology of releasing mechanisms: prey-catching in toads. *Behavioral and Brain Sciences* **10**: 337–403.

Feng, A. S., Simmons, J. A. & Kick, S. A. (1978) Echo detection and target-ranging neurons in the auditory system of the bat *Eptesicus fuscus*. *Science* **202**: 645–648.

Fisher, R. A. (1958) *The genetical theory of natural selection.* 2nd edn. New York: Dover.

Gallistel, C. R. (1990) *The organization of learning*. Cambridge: MIT Press.

Gould, J. L. & Gould, C. G. (1989) *Sexual selection.* New York: W. H. Freeman.

Gould, J. L. & Marler, P. (1987) Learning by instinct. *Scientific American* **256**(1): 62–73.

Gould, S. J. (1978) *Ever since Darwin*. London: Burnett.

Gould, S. J. & Lewontin, R. C. (1979) The spandrels of San Marco and the Panglossian paradigm; a critique of the adaptationist programme. *Proceedings of the Royal Society of London B* **205**: 581–598.

Grafen, A. (1982) How not to measure inclusive fitness. *Nature, London* **298**: 425–426.

Grafen, A. (1984) Natural selection, kin selection and group selection. In: *Behavioural ecology. An evolutionary approach.* 2nd edn. (eds. J. R. Krebs & N. B. Davies). Oxford: Blackwell Scientific Publications. pp. 62–84.

Grafen, A. (1990a) Biological signals as handicaps. *Journal of Theoretical Biology* **144**: 517–546.

Grafen, A. (1990b) Sexual selection unhandicapped by the Fisher process. *Journal of Theoretical Biology* **144**: 473–516.

Grafen, A. (1990c) Do animals really recognize kin? *Animal Behaviour* **39**: 42–54.

Grafen, A. (1991) Modelling in behavioural ecology. In: *Behavioural ecology. An evolutionary approach.* 3rd edn. (eds J. R. Krebs & N. B. Davies) Oxford: Blackwell Scientific Publications. pp. 5–31.

Greenberg, L. (1979) Genetic component of bee odor in kin recognition. *Science* **206**: 1095–1097.

Griffin, D. R. (1976) *The question of animal awareness.* New York: Rockefeller University Press.

Griffin, D. R. (1993) *Animal minds*. Chicago: University of Chicago Press.

Guilford, T. & Dawkins, M. S. (1991) Receiver psychology and the evolution of animal signals. *Animal Behaviour* **42**: 1–14.

Guilford, T. & Dawkins, M. S. (1993) Receiver psychology and the design of animal signals. *Trends in NeuroSciences* **16**: 430–436.

Hailman, J. P. (1967) How an instinct is learned. *Scientific American* **221**(6): 98–108.

Halliday, T. R. & Slater, P. J. B. eds. (1983) *Communication*. Vol. 2 of *Animal Behaviour*. Oxford: Blackwell Scientific Publications.

Hamer, D. H., Hu, S., Magnuson, V. L., Hu, N. & Pattatucci, A. M. L. (1993) A linkage between DNA markers on the X chromosome and male sexual orientation. *Science* **261**: 321–327.

Hamilton, W. D. (1964) The genetical evolution of social behaviour. *Journal of Theoretical Biology* 7: 1–52.
Hamilton, W.D. (1980) Sex versus non-sex versus parasite. *Oikos* **35**: 282–290.
Hamilton, W. D., Axelrod, R. & Tanses, R. (1990) Sexual reproduction as an adaptation to resist parasites (a review). *Proceeding of the National Academy of Sciences* **87**: 3566–3573.
Hamilton, W. D. & Zuk, M. (1982) Heritable true fitness in birds: a role for parasites? *Science* **218**: 384–386.
Harvey, P. H. (1985) When the going gets tough. *Nature* **317**: 388–389.
Harvey, P. H. & Pagel, M. D. (1991) T*he comparative method in evolutionary biology*. Oxford: Oxford University Press.
Harvey, P. H. & Ralls, K. (1986) Do animals avoid incest? *Nature Lond.* **320**: 575–576.
Henderson, N. D. (1970) Genetic influences on the behaviour of mice can be obscured by laboratory rearing. *Journal of Comparative and Physiological Psychology* **72**: 505–511.
Herrnstein, R. J. (1990) Levels of stimulus control: a functional approach. *Cognition* **37**: 133–166.
Hert, E. (1989) The function of egg-spots in an African mouth-brooding cichlid fish. *Animal Behaviour* **37**: 726–732.
Heyes, C. M. (1993) Anecdotes, training, trapping and triangulating: do animals attribute mental states? *Animal Behaviour* **46**: 177–188.
Hinde, R. A. (1960) Energy models of motivation. *Symposium of the Society for Experimental Biology* **14**: 190–213.
Hinton, G. E. (1992) How neural networks learn from experience. *Scientific American* **267** (3): 145–151.
Hogan, J. A. (1989) The interaction of incubation and feeding in broody junglefowl hens. *Animal Behaviour* **38**: 121–128.
Hoogland, J. L. & Sherman, P. W. (1976) Advantages and disadvanatges of Bank swallow coloniality. *Ecological Monographs* **46**: 33–58.
Houston, A. I. & McNamara, J. M. (1990) Risk-sensitive foraging and temperature. *Trends in Ecology and Evolution* **5**: 131–132.
Hsia, D. Y.-Y. (1970) Phenylketonuria and its variants. In: *Progress in Medical Genetics*, vol. 7 (eds A. G. Steinberg & A. G. Bearn) New York: Grune & Stratton, pp. 29–68.
Humphrey, N. K. (1983) *Consciousness regained.* Oxford University Press.
Huxley, J. (1966) Ritualisation of behaviour in animals and men. *Philosophical Transactions of the Royal Society of London.* B **251**: 249–271.
Jacob, F. (1977) Evolution and tinkering. *Science* **196**: 1161.
Jarvis, J. U. M.(1981) Eusociality in a mammal: cooperative breeding in naked mole rat colonies. *Science* **212**: 571–573.
John, E. R., Tang, Y., Brill, A. B., Young, R. & Ono, K. (1986) Double-labelled metabolic maps of memory. *Science* **233**: 1167–1175.
Kettlewell, H. B. D. (1955) Selection experiments on industrial melanism in the Lepidoptera. *Heredity* **9**: 323–335.
Kimura, M. (1991) Recent developments of the neutral theory viewed from the Wrightian tradition of theoretical population genetics. *Proceedings of the National Academy of Sciences* **88**: 5969–5973.
Kirkpatrick, M. (1982) Sexual selection and the evolution of female choice. *Evolution* **36**: 1–12.

Kodric-Brown, A. (1992) Male dominance can enhance mating success in guppies. *Animal Behaviour* **44**: 165–167.

Konishi, M. (1985) Bird song: from behaviour to neuron. *Advances in Neuroscience* **8**: 125–170.

Konishi, M. (1986) Centrally synthesized maps of sensory space. *Trends in NeuroSciences* **9**: 163–168.

Krebs, J. R. & Davies, N. B. (1981) *An introduction to behavioural ecology*. 1st edn. Oxford: Blackwell Scientific Publications.

Krebs, J. R. & Davies, N. B. (1993) *An introduction to behavioural ecology*. 3rd edn. Oxford: Blackwell Scientific Publications.

Krebs, J. R. & Kacelnik, A. (1991) Decision-making. In *Behavioural ecology* (eds J. R. Krebs & N. B. Davies. 3rd edn. Oxford: Blackwells Scientific, pp. 105–136.

Lande, R. (1982) Rapid origin of sexual isolation and character divergence within a cline. *Evolution* **36**: 213–223.

Le Boeuf, B. J. (1972) Sexual behaviour in the northern elephant seal, *Mirounga angustirostris*. *Behaviour* **41**: 1–26.

Lehrman, D. (1953) A critique of Konrad Lorenz's theory of instinctive behavior. *Quarterly Review of Biology* **28**: 337–363.

Lemon, W. C. (1991) Fitness consequences of foraging behaviour in the zebra finch. *Nature Lond.* **352**: 153–155.

Lewontin, R. C., Rose, S., & Kamin, L. J. (1984) *Not in our genes*. Harmondsworth: Penguin.

Lockery, S. R. & Sejnowski, T. J. (1993) The computational leech. *Trends in NeuroSciences* **16**(7): 283–290.

Lockery, S. R., Wittenberg, G., Kristau, W. B. & Cottrell, G. W. (1989) Function of identified interneurones in the leech elucidated using neural networks trained by back-propagation. *Nature, Lond.* 340: 468-471.

Lorenz, K. (1932) Betrachtungen über das Erkennen der arteigenen Triebhandlungen der Vogel. (A consideration of methods of identification of species-specific instinctive behaviour patterns in birds). *Journal für Ornithologie* **80** (1). Translated by R. Martin in K. Lorenz (1970) *Studies in animal and human behaviour* Vol. 1. London: Methuen. pp. 57–100.

Lorenz, K. (1937) Uber die Bildung des Instinktbegriffes (The establishment of the instinct concept). *Die Naturwissenschaften* **25** (19), 289–300, 307–318, 324–331. Translated by R. Martin in K. Lorenz (1970) *Studies in animal and human behaviour* Vol. 1. London: Methuen, pp. 259–315.

Lorenz, K. (1950) The comparative method in studying innate behaviour patterns. *Symposium of the Society for Experimental Biology* **4**: 221–268.

Lorenz, K. (1965) *Evolution and modification of behaviour*. Chicago: University of Chicago Press.

Lythgoe, J. N. (1979) *The ecology of vision*. Oxford: Clarendon Press.

McClean, I. G. & Rhodes, G. (1991) Enemy recognition and response in birds. *Current Ornithology* **8**: 173–211. (ed. D. M. Power). New York: Plenum.

McFarland, D. (1989) *Problems of animal behaviour*. Harlow: Longman.

McFarland, D. J. & Houston, A. I. (1981) *Quantitative ethology. The state space approach*. London: Pitman.

McGrew, W. C. (1992) *Chimpanzee material culture*. Cambridge: Cambridge University Press.

McNeill Alexander, R. (1982) *Optima for animals*. London: Edward Arnold.

Magnus, D. (1958) Experimentelle Untersuchungen zur Bionomie und Ethologie des Kaisermantels *Argynnis paphia* L. (Lep. Nymph): I Uber optische Auslöser von Anfliergereaktionen und ihre Bedeutung für das Sichfinden der Geschlechter. *Zeitschrift für Tierpsychologie* **15**: 397–426.

Magurran, A. E., Seghers, B. H., Carvalho, G. R. & Shaw, P. W. (1992) Behavioural consequences of an artificial introduction of guppies (*Poecilia reticulata*) in N.Trinidad: evidence for the evolution of anti-predator behaviour in the wild. *Proc. Roy. Soc. Lond. B* **248**: 117–122.

Martin, R. D. (1984) Scaling effects and adaptive strategies in mammalian lactation. *Symp. Zool. Soc. Lond.* **51**: 81–117.

Mason, G. (1991) Stereotypes: a critical review. *Animal Behaviour* **41**: 1015–1038.

Maynard Smith, J. (1976) Sexual selection and the handicap principle. *Journal of Theoretical Biology* **57**: 239–242.

Maynard Smith, J. (1978) Optimization theory in evolution. *Annual Review of Ecology and Systematics* **9**: 31–56.

Maynard Smith, J. (1991) Theories of sexual selection. *Trends in Ecology and Evolution* **6**(5): 146–151.

Maynard Smith, J. & Harper, D. G. C. (1988) The evolution of aggression: can selection generate variability? *Philosophical Transactions of the Royal Society of London* B **319**: 557–570.

Medin, D. L. & Smith, E. E. (1984) Concepts and concept formation. *Annual review of psychology* **35**: 113–138.

Miller, N. E. (1957) Experiments on motivation. Studies combining psychological, physiological and pharmacological techniques. *Science* **126**: 1271–1278.

Mock, D. W. (1984) Siblicidal aggression and resource monopolization in birds. *Science* **225**: 731–733.

Møller, A. (1987) Social control of deception among status signalling house sparrows *Passer domesticus*. *Behavioural Ecology and Sociobiology* **20**: 307–311.

Morgan, T. H. (1911) Random segregation versus coupling in Mendelian inheritance. *Science* **33**: 384.

Nelson, J. B. (1967) Colonial and cliff-nesting in the gannet. *Ardea* **55**: 60–90.

Oakley, D. A. ed. (1985) *Brain and mind*. London: Methuen.

Oatley, K. (1978) *Perceptions and representations*. London: Methuen.

Ornstein, R. (1991) *The evolution of consciousness*. New York: Simon & Schuster.

O'Shea, M. & Williams, J. L. D. (1974) The anatomy and output connection of locust visual interneurone; the lobular giant movement detector (LGMD) neurone. *Journal of Comparative Physiology* **91**: 257–266.

Papi, F. ed. (1992) *Animal homing*. London: Chapman & Hall.

Parker, G. A. (1982) Phenotype-limited evolutionarily stable strategies. In: *Current problems in sociobiology* (ed. King's College Sociobiology Group). Cambridge: Cambridge University Press, pp. 173–201.

Parker, G. A. & Maynard Smith, J. (1990) Optimality theory in evolutionary biology. *Nature* **348**: 27–33.

Partridge, B. L. & Pitcher, T. J. (1979) Evidence against a hydrodynamic function for fish schools. *Nature, Lond.* **279**: 418–419.

Patterson, I. J. (1965) Timing and spacing of broods in the black-headed gull *Larus ridibundus*. *Ibis* **107**: 433–459.

Pepperberg, I. (1987) Evidence for conceptual qualitative abilities in the African grey parrot (*Psittacus erithacus*). *Zeitschrift für Tierpsychologie* **68**: 89–114.

Perdeck, A. C. (1958) Two types of orientation in migrating starlings, *Sturnus vulgaris* L., and chaffinches, *Fringilla coelebs L.*, as revealed by displacement experiments. *Ardea* **46**: 1–37.

Pérusse, D. (1993) Cultural and reproductive success in industrial societies: testing the relationship at the proximate and ultimate levels. *Behavioral and Brain Sciences* **16**: 267–332.

Pinker, S. & Bloom, P. (1992) Natural language and natural selection. In: *The adapted mind* (eds J. H. Barkow, L. Cosmides & J. Tooby). New York: Oxford University Press, pp. 451–494.

Pomiankowski, A., Iwasa, Y. & Nee, S. (1991) The evolution of costly mate preferences. 1. Fisher and biased mutation. *Evolution* **45**: 1422–1430.

Potts, W. K. (1984) The chorus-line hypothesis of manoeuvre coordination in avian flocks. *Nature Lond.*, **309**: 344–345.

Premack, D. (1983) Animal cognition. *Annual Review of Psychology* **34**: 351–362.

Pusey, A. (1980) Inbreeding avoidance in chimpanzees. *Animal Behaviour* **28**: 575–576.

Quinlan, P. (1991) *Connectionism in psychology*. New York: Harvester Wheatsheaf.

Ridley, M. (1983) *The explanation of organic diversity: the comparative method and adaptations for mating.* Oxford: Oxford University Press.

Roitblat, H. L. & von Ferson (1992) Comparative cognition: representation and processes in learning and memory. *Annual Review of Psychology* **43**: 671–710.

Romer, H. (1993) Environmental and biological constraints for the evolution of long-range signalling and hearing in acoustic insects. *Philosophical Transactions of the Royal Society of London* B *340*: 187–195.

Rubenstein, D. I. (1980) On the evolution of alternative mating strategies. In: *Limits to action. The allocation of individual behavior.* (ed. J. Staddon). New York: Academic Press, pp. 65–100.

Rushen, G. (1991) Problems associated with the interpretation of physiological data in the assessment of animal welfare. *Applied Animal Behaviour Science* **28**: 381–386.

Ryan, M. J. & Rand, A. S. (1993) Sexual selection and signal evolution: the ghost of biases past. *Philosophical Transactions of the Royal Society of London* **340**: 187–195.

Sales, G. & Pye, D. (1974) *Ultrasonic communication by animals*. London: Chapman & Hall.

Shannon, C. E. & Weaver, W. (1949) *The mathematical theory of communication*. Urbana: University of Illinois Press.

Shepher, J. (1971) Mate selection among 2nd generation kibbutz adolescents and adults: incest avoidance and negative imprinting. *Archives of Sexual Behavior* **1**: 293–307.

Sherrington, C. S. (1906) *The integrative action of the nervous system*. New York: Scribner's.

Sherry, D. F., Mrosovsky, N. & Hogan, J. A. (1980) Weight loss and anorexia during incubation in birds. *Journal of Comparative and Physiological Psychology* **94**: 89–98.

Simmons, J. A., Fenton, M. B. & O'Farrell, M. J. (1979) Echolocation and the pursuit of prey in bats. *Science* **203**: 16–21.
Simmons, J. A. & Vernon, J. A. (1971) Echolocation: discrimination of targets by the bats *Eptesicus fuscus. Journal of Experimental Zoology* **176**: 315–328.
Simpson, S. J. & Ludlow, A. R. (1986) Why locusts start to feed: a comparison of causal factors. *Animal Behaviour* **34**: 480–496.
Slater, P. J. B. (1989) Bird song learning: causes and consequences. *Ethology, Ecology and Evolution* **1**: 19–46.
Sturtevant, A. H. (1913) The linear arrangement of six sex-linked factors in *Drosophila*, as shown by their mode of association. *Journal of Experimental Zoology* **14**: 43–59.
Symons, D. (1992) On the use and misuse of Darwinism in the study of human behavior. In: *The adapted mind* (eds. J. H. Barkow, L. Cosmides & J. Tooby). New York: Oxford University Press, pp. 137–159.
Tanake, H., Wong, D. & Taniguchi, I. (1992) The influence of stimulus duration on the delay timing of cortical neurons in the FM bat *Myotis lucifugus. Journal of Comparative Physiology A* **171**: 29–40.
Thornhill, R. (1976) Sexual selection and nuptial feeding behaviour in *Bittacus apicalis* (Insecta: Mecoptera). *American Naturalist* **110**: 529–548.
Thorpe, W. H. (1961) *Bird song.* Cambridge: Cambridge University Press.
Tinbergen, N. (1932) On the orientation of the Digger Wasp *Philanthus triangulum* Fabr. Reprinted in *The animal in its world* (1972). London: Allen & Unwin, pp. 103–127.
Tinbergen, N. (1951) *The study of instinct.* Oxford: Oxford University Press.
Tinbergen, N. (1952) 'Derived' activities, their causation, biological significance, origin and emancipation during evolution. *Quarterly Review of Biology* **27**: 1–32.
Tinbergen, N. (1963) On aims and methods of ethology. *Zeitschrift für Tierpsychologie* **20**: 410–433.
Tinbergen, N., Broekhuysen, G. J., Feekes, F., Houghton, J. C. W., Kruuk, H. & Szulc, E. (1967) Egg shell removal by the black-headed gull, *Larus ridibundus* L.: a behaviour component of camouflage. *Behaviour* **19**: 74–117.
Toates, F. (1986) *Motivational systems.* Cambridge: Cambridge University Press.
Tolman, E. C. (1932) *Purposive behavior in animals and men.* New York: Century.
Tooby, J. & Cosmides, L. (1992) The psychological foundations of culture. In: *The adapted mind.* (eds J. H. Berkow, L. Cosmides & J. Tooby). New York: Oxford University Press, pp. 19–136.
Trivers, R. L. (1971) The evolution of reciprocal altruism. *Quarterly Review of Biology* **46**: 35–57.
Tullock, G. (1971) The coal tit as a careful shopper. *American Naturalist* **105**: 77–79.
Van den Burghe, P. (1983) Human inbreeding avoidance. Culture in nature. *Behavioral and Brain Sciences* **6**: 9–124.
Van Valen, L. (1973) A new evolutionary law. *Evolutionary Theory* **1**: 1–30.
Vaughan, W. Jr. & Green, S. L. (1984) Pigeon visual memory capacity. Journal of Ex*perimental Psychology Animal Behavior Processes* **10**: 256–271.
Velman, M. (1991) Is human information processing conscious? *Behavioral*

and Brain Sciences **14**: 651–726.

Von Ferson, L. & Lea, S. E. G. (1990) Category discrimination of polymorphous stimulus sets by pigeons. *Quarterly Journal of Experimental Psychology* **30**: 521–537.

Weihs, D. (1975) Some hydrodynamical aspects of fish schooling. In: *Swimming and flying in nature* (eds. T. Y.-T. Wu, C. J. Brockaw and C. Brennan). New York: Plenum Press, pp. 703–18.

Wickler, W. (1962) 'Egg-dummies' as natural releasers in mouth-breeding cichlids. *Nature, Lond.* **194**: 1092–1094.

Wiley, R. H. (1993) Errors, exaggeration and deception in animal communication. In: *Behavioral mechanisms in ecology* (ed. L. Real). Chicago: University of Chicago Press, in press.

Williams, G. C. (1975) *Sex and evolution.* Princeton: Princeton University Press.

Williams, G. C. (1992) *Natural selection: domains, levels and challenges.* Oxford: Oxford University Press.

Wilson, M. & Daly, M. (1992) The man who mistook his wife for a chattel. In: *The adapted mind* (eds J. H. Barkow, L. Cosmides & J. Tooby). New York: Oxford University Press, pp. 289–322.

Wright, S. (1932) The roles of mutation, inbreeding, crossbreeding, and selection in evolution. *Genetics* **1**: 356–366.

Young, D. (1989) *Nerve cells and animal behaviour.* Cambridge: Cambridge University Press.

Ydenberg, R. C. & Houston, A. I. (1986) Optimal trade-off between competing behavioural demands in the great tit. *Animal Behaviour* **34**: 1041–1050.

Zahavi, A. (1975) Mate selection – a selection for a handicap. *Journal of Theoretical Biology* **53**: 205–214.

Zahavi, A. (1977) Reliability in communication systems and the evolution of altruism. In: *Evolutionary ecology* (eds B. Stonehouse and C. M. Perrins). London: Macmillan, pp. 253–259.

Subject index

adaptation, 1–17, 27, 108
  and genetics, 56, 57
  human, 160
adaptive landscape, 35
alarm call, 73
angler fish, 79
ants, 45, 49
artificial mutants, 8–9
asexual reproduction, 90, 92

baboons, 154
badges of status, 95
barn owls, 120
bats, echo location in, 12–14, 17–18, 27, 120
bees
  dance, 114
  sterile castes, 45, 49
behavioural ecology, 108, 120
big-mouth gene, 51, 52
black boxes, 112, 114, 115, 119
bluehead wrasse, 89
bowerbird, 95
brother-helping, *see* sibling care
butterflies, 102

chaffinch, 65
cheat, 81, 85, 100
chimpanzee
  culture, 156, 157, 162
  inbreeding avoidance, 154
  mating system, 160
choice, 81
cichlid fish, 86, 88
circuit diagrams, 113, 114, 119
coefficient of relatedness, *see* relatedness
cognitive
  definition, 123–5, 138
  maps, 127, 131
  representations, 123–5, 132, 134, 139, 142, 144
Comparative Method, 9, 10
competition, 118
concept formation, 131–134
connectionism, 119
consciousness
  distinction from cognitive, 123–125, 135
  problem of, 139–140, 146–7
  working definition, 137–138
cost function, 33, 34
crickets, genetics of, 58–59, 63–67
culture, 153–157, 161–163

design features, 12–15, 26, 28, 30
determinism, 149, 151
developmentally fixed, 63–64, 67–68
digger wasp, 127
displacement activities, 111
display, 74
dunnock, 162

echo location, 12, 18, 120
economics, 32
efficacy, 103, 104
egg-spots, 88
eggshell removal, 8–12, 56, 107, 160
electric fish, 115
elephant seals, 92, 93, 94
elephants, 37
epiphenomenalism, 140
exploitation, 10, 92, 94

female choice, 94, 97, 98, 99, 153
fish school, 16
fitness, 37, 42
  inclusive, 2, 38–44
flocking in birds, 72
free will, 152
fruitflies, 38, 57

gannet, 11, 12
genetic determinism, 60, 61, 68, 150, 152, 162
genetic drift, 18
gibbons, 160
goal function, 33, 34
'good genes', 99, 100, 101, 104
gorilla, 149, 160, 161
great tits, 25
green beard genes, 51, 52
group living, 6
gulls
  black-headed, 5, 8, 56, 107, 160
  herring, 124
  laughing, 65, 66
guppies, 7, 94

handicap principle, 80, 81, 82, 83, 85, 86, 99, 100, 101, 102
haplodiploid genetics, 49
homosexuality, gene for, 55, 150
honest signalling, 80-87, 96, 104
Hymenoptera, 49

inbreeding depression, 153–4
information transfer, 75–78, 86–88, 104
innate behaviour, 56, 61–69, 111, 124
  schoolmarm, 66
instinct, 61, 109–111, 120

jumping spiders, 88
junglefowl, 71, 118
Just-so stories, 2, 17

kittiwake, 10, 11, 12, *see also* gulls

language, humans, 64, 156
learning, 63, 65–66
leech, 113, 120
locust, 113, 116

male–male competition, 95
manipulation, 78, 79, 85, 87, 88, 103
mantis shrimp, 85
map, cognitive, 127, 130
matching-to-sample, 133
mate choice, 154-5, *see also* sexual selection
maze-learning, 60, 61
meerkats, 21
mental processes, 123
mice, 61
mimicry, 79, 85
monogamy, 162
motivation, 109, 114, 115, 116, 117, 118, 124, 144, 145, 146
mutations 3, 9, 18, 22, 39, 158

naked mole rat, 45
navigation, 128, 129, 130
neural networks, 109, 119, 120
neurobiology, 108, 112–114, 117–120
neurons, 112, 113

nightjars, 12, 160
nothing but, 165
number, concept of, 131

Optimal Foraging Theory, 24, 26, 28–29, 30, 117
optimal, 21
  constraints on optimality, 29
orang-utans, 160
orientation, 128, 129, 130

pain, 144, 148
parasites, 23, 91, 100
parrots, 134
passive attraction, 101
pea plants, genetics of, 57
peppered moth, 5, 6
phenylketonuria, *see* PKU
pigeon, 123, 128, 132, 134, 140
pilotage, 128, 129, 130
PKU, 60, 150, 151
polymorphous features, 133, 134

relatedness, degree of, 42, 49, 50
relatedness, coefficient of, 38, 41–42, 47–51
racoons, 160
rats, 45, 116, 127, 153, 156
receiver psychology, 88
red deer, 73, 78, 79, 83, 93, 114, 143
Red Queen hypothesis, 23, 33
reflexes, 109, 143, 144
relatives
  helping, 37, 39, 44–45, 48–49, 52–53, 57
  recognizing, 59
representation, cognitive, 125, 126, 135
ritualization, 72, 75, 80, 104
rules of thumb, 31, 35
runaway theory, 97–98

sage grouse, 92
satellite males, 58
schooling in fish, 15, 72
sea-slugs, 113 , 120
seals, 161
selection
  artificial, 9
  compromise 29
sensory bias, 103, 104
sensory exploitation, 102, 103
sex, 89
sexual selection, 94, 97, 108
sexual signals, 88, 101, 103
sib competition, 91
siblicide, 50
sibling care, 48, 49, 51, 57
sign stimuli, 108
signal, 72–79, 81–88, 96, 100, 104
simple weighted sum, 42, 43, 46
social intellect theory, 141
social insects, 57
sociobiology, 108
song sparrow, 66
sows, 145
spandrels, 18
starlings, 129
status badges, 85, 86, 95, 96
stereotypes, 146
sterile castes, 45
stick insects, 160
stickleback, 63, 64, 65, 160
suffering, 138, 144, 145
supernormal stimuli, 102, 108
swan, 107
sweat bees, 59

termites, 48, 156
time lags, 24, 91, 157, 158, 159, 163
toads, 101, 120, 158
tree-shrews, 161
tungara frog, 102

utility, 32

vacuum activities, 111
vervet monkeys, 126, 135

von Ferson, 131, 133

wasps, 45, 49

Author index

Abraham, M.V., 29, 167
Adams, E.S., 85, 167
Alatalo, R., 96, 167
Albon, S., 73, 78, 143, 167
Altman, J.S., 113, 167
Andersson, M., 81, 167
Arak, A., 101, 167
Arey, D.S., 145, 167
Austad, S.N., 48, 167

Baars, B., 141, 167
Baker, R.R., 128, 130, 131, 167
Balmford, A., 101, 167
Barnard, C., 52, 53, 167
Bateson, P.P.G., 97, 109, 147, 167
Bekoff, M., 123, 167
Bell, G., 23, 91, 167
Ben Shaul, D.M., 161, 167
Bentley, D., 58, 64, 168
Bloom, P., 156, 174
Bodmer, W., 153, 168
Bonner, J.T., 153, 168
Borgia, G., 95, 168
Brenner, S., 90
Burrows, M., 113, 168
Buss, D.M.,155, 168
Byrne, R., 135, 168

Cade, W., 58, 168
Caldwell, R. L., 85, 167
Caraco, T., 117, 168
Cavalli-Sforza, L.L., 153, 168
Cheney, D.L., 126, 135, 168
Cheng, K. 131, 168
Clark, D.L., 88, 168
Clutton-Brock, T.H., 73, 78, 93, 105, 143, 161, 168
Colgan, P., 120, 168
Collett, T.S., 131, 168
Coopersmith, C.B., 94, 168
Cosmides, J., 124, 158, 163, 175
Cowie, R., 25, 168
Cullen, E., 10, 11, 169
Cullen, J.M., 73, 169

Daan, S., 27, 169
Daly, M., 162, 176
Darwin, C., 2, 92, 94, 96, 100, 109, 169
Davies, N.B., 42, 74, 82, 162, 169
Davis, J.M., 72, 134, 169
Dawkins, M.S., 88, 103, 147, 169
Dawkins, R., 18, 19, 20, 24, 35, 51, 53, 69, 73, 78, 87, 150, 169
Dennett, D.C., 142, 147, 169
Drent, R.H., 27, 169
Durham, W.H., 163, 169

Edwards, C.A., 131, 169
Elgar, M., 6, 117, 169
Ellis, B.J., 155, 169
Emlen, S.T., 41, 169

Endler, J. A., 19, 87, 169, 170
Ewert, J.-P. 120, 170

Fisher, R.A., 93, 97, 98, 99, 100, 101, 102, 104, 170

Gallistel, C.R., 125, 127, 130, 170
Gould, S.J., 2, 17, 18, 19, 20, 55, 69, 105, 170
Grafen, A., 37, 42, 43, 46, 52, 53, 81, 82, 84, 170
Green, S.L., 132, 175
Greenberg, L., 59, 170
Griffin, D.R., 123, 135, 137, 138, 147, 170
Guilford, T., 88, 103, 170

Hailman, J.P., 65, 69, 170
Halliday, T.R., 82, 88, 170
Hamer, D.H., 55, 170
Hamilton, W.D., 37, 38, 39, 40, 41, 46, 47, 50, 51, 53, 91, 92, 100, 109, 171
Harper, D.G.C., 85, 95, 173
Harvey, P.H., 12, 20, 30, 154, 171
Herrnstein, R.J., 131, 171
Hert, E., 88, 171
Heyes, C.M., 135, 171
Hinde, R.A., 111, 171
Hinton, G.E., 120, 171
Hogan, J.A., 118, 171
Honig, W.K., 131, 169
Hoogland, J.L., 20, 171
Houston, A.I., 29, 30, 35, 143, 171, 176
Hoy, R., 58, 64, 168
Hsia, D.Y.-Y., 60, 171
Humphrey, N.K., 141, 171
Huxley, J., 72, 171

Jacob, F., 34, 171
Jarvis, J.U., 45, 171

Kacelnik, A., 26, 30, 35, 172
Kein, J., 113, 167
Kettlewell, H.B.D., 5, 171
Kimura, M., 18, 171
Kirkpatrick, M., 97, 98, 171
Kodric-Brown, A., 94, 172
Konishi, M., 120, 172
Krebs, J.R., 26, 30, 35, 42, 73, 78, 87, 172

Lande, R., 97, 98, 172
Le Boeuf, B.J., 95, 172
Lea, S.E.G., 133, 176
Lehrman, D., 63, 172
Lemon, W.C., 27, 172
Lenington, S., 94, 168
Lewontin, R.C., 2, 17, 18, 19, 20, 55, 172
Lockery, S.R., 113, 172
Lorenz, K., 62, 64, 65, 68, 109, 110, 111, 112, 114, 120, 172
Ludlow, A.R., 116, 175
Lythgoe, J.N., 87, 172

McClean, I.G., 123, 126, 172
McFarland, D.J., 33, 123, 143, 172
McGrew, W.C., 153, 172
McNamara, J., 30, 171
McNeill Alexander, R., 35, 172
Magnus, D., 102, 173
Magurran, A., 7, 173
Marler, P., 69, 170
Martin, R.D., 161, 173
Mason, G., 146, 173
Maynard Smith, J., 24, 35, 80, 82, 85, 95, 99, 105, 173
Medin, D.L., 131, 173
Miller, N.E., 117, 173
Mock, D.W., 50, 173
Møller, A., 86, 95, 96, 173
Morgan, T.H., 115, 173

Nelson, J.B., 11, 173

O'Shea, M., 113, 173
Oakley, D.A., 173
Oatley, K., 114, 173
Ornstein, R., 142, 173

Pagel, M., 12, 20, 171
Papi, F., 129, 130, 173
Parker, G.A., 35, 98, 101, 173
Partridge, B.L., 15, 173
Patterson, I.J., 5, 173
Pepperberg, I., 134, 174
Perdeck, A.C., 129, 174
Pérusse, D., 134, 155
Pinker, S. 156, 174
Pitcher, T.J., 15, 173
Pomiankowski, A., 99, 174
Potts, W.K., 72, 174
Premack, D., 133, 174
Pusey, A., 154, 174
Pye, D., 13, 174

Queller, 51
Quinlan, P., 119, 174

Rabenold, K.N., 48, 167
Ralls, K., 154, 171
Rand, A.S., 102, 174
Rhodes, G., 123, 126, 172
Ridley, M., 12, 174
Roitblat, H.L., 131, 133, 174
Romer, H., 87, 174
Rose, S., 69, 172
Rowell, C.H.F., 113, 168
Rushen, G., 146, 174
Ryan, M.J., 102, 174

Sales, G., 13, 174
Sejnowski, T.J., 113, 172
Seyfarth, R.M., 126, 135, 168
Shannon, C.E., 76, 174
Shepher, J., 154, 174
Sherman, P., 20, 171
Sherrington, C.S., 143, 174
Sherry, D.F., 118, 174
Simmons, J.A., 13, 120, 175
Simpson, S.J., 116, 175
Slater, P.J.B., 88, 175
Smith, E.E., 131, 173
Sturtevant, A.H., 115, 175
Symons, D., 156, 163, 175

Tanake, H., 13, 175
Thornhill, R., 96, 175
Thorpe, W.H., 65, 175
Tinbergen, N., 8, 20, 32, 72, 74, 87, 102, 107–112, 114, 120, 127
Toates, F., 117, 175
Tolman, E.C., 127, 175
Tooby, J., 124, 158, 163, 175
Trivers, R.L., 109, 175
Tullock, G., 32, 175

Uetz, G.W., 88, 168

Van den Burghe, P., 154, 175
Van Valen, L., 23, 175
Vaughan, W., Jr., 132, 175
Velman, M., 138, 175
Vernon, J.A., 120, 175

Weaver, W., 76, 174
Whiten, A., 135, 168
Wickler, W., 88, 176
Wiley, R.H., 85, 176
Williams, G.C., 19, 90, 113, 156, 157, 176
Wilson, M., 162, 176
Wrege, P.H., 41, 169
Wright, S., 18, 34, 176

Ydenberg, R., 29, 35, 176
Young, D., 120, 176
Zahavi, A., 80, 81, 84, 99, 176
Zuk, M., 100, 176